Mapping Census 2010

The Geography of American Change

Riley Peake

Esri Press
REDLANDS|CALIFORNIA

Esri Press, 380 New York Street, Redlands, California 92373-8100
Copyright © 2012 Esri

Ask for Esri Press titles at your local bookstore or order by calling 800-447-9778, or shop online at esri.com/esripress. Outside the United States, contact your local Esri distributor or shop online at eurospanbookstore.com/esri.

Esri Press titles are distributed to the trade by the following:

In North America
Ingram Publisher Services
Toll-free telephone: 800-648-3104
Toll-free fax: 800-838-1149
E-mail: customerservice@ingrampublisherservices.com

In the United Kingdom, Europe, Middle East and Africa, Asia, and Australia
Eurospan Group
3 Henrietta Street
London WC2E 8LU
United Kingdom
Telephone: 44(0) 1767 604972
Fax: 44(0) 1767 601640
E-mail: eurospan@turpin-distribution.com

Contents

Section 1

Location Maps and Total Population

Section 2

Households

Housing

Section 3

Diversity

Section 4

Non-Hispanic White

Black or African American

American Indian or Alaska Native

Asian

Native Hawaiian or Other Pacific Islander

Two or More Races

Hispanic or Latino Origin

Preface

The recent decennial survey conducted by the US Census Bureau was an enormous undertaking that generated a wealth of information about the state of the nation and its citizens. When charts and tables, figures and numbers, take on locations and are mapped with skill using GIS software such as ArcGIS for Desktop, we can see patterns and change, and potential associations within. Demographic data is immensely powerful and eye-opening, especially when transformed into knowledge via a map.

The ten years since the previous census has been a decade fraught with unease in global security, natural disasters, and economic turmoil. How has America *really* changed demographically and geographically in these difficult times? *Mapping Census 2010: The Geography of American Change* sheds light on this and other questions about our population. Coupling the hard work of the Census Bureau with the latest Esri technology, this book presents beautiful and engaging maps that show us trends in age, race, housing, households, and movement within each demographic.

I hope that this work will inspire you to seek your own answers in the census data. Anyone can go to ArcGIS Online and analyze US Census data using maps. I encourage you to learn more about the states of housing and population in America. But first, take some time to study these pages. I think you'll appreciate how the maps show the story of our nation's diversity, resilience, and change.

Sincerely,

Jack Dangermond
President, Esri

Introduction

Mapping Census 2010: The Geography of American Change depicts the extraordinary changes in the American population over the past decade. Census 2010 counts reflect the changes through new measures like multigenerational households. Comparisons to Census 2000 counts provide a measure of perspective. These maps show the distribution of change across the country.

The population increased by almost 10 percent from 2000 to 2010, but total population change masks the variation in growth by race or Hispanic origin—or by region. The majority population, non-Hispanic white, increased in 2010, but accounted for less than 10 percent of total change. In the Northeast and Midwest, the majority population *decreased* from 2000 to 2010. All of the population growth in these regions was contributed by minority populations.

The Hispanic population accounted for over half of the change in the US population from 2000 to 2010 with an increase of 43 percent. Regionally, the increase in the Hispanic population ranged from 33 percent in the Northeast to 57 percent in the South. Growth among the Asian populations was comparable, 43 percent, and the regional variation was also evident, from 33 percent in the West to 67 percent in the South. The South experienced the most rapid growth in minority populations—and in the total population. The effect of these changes was the largest increase in population diversity in the South.

Esri developed an index in 2000 to reflect the racial and ethnic diversity of the population. Our Diversity Index represents the likelihood that two persons, chosen at random from the same area, belong to different race or ethnic groups. If an area's entire population belongs to one race group and one ethnic group, then the area has zero diversity. An area's diversity index increases to 100 when the population is evenly divided into two or more race/ethnic groups. The 2010 Diversity Index for the United States stands at 61, which means the probability that two people randomly chosen from the US population would belong to different race or ethnic groups is 61 percent. The Diversity Index varies from a low of 41 in the Midwest to a high of 73 in the West. The differences are evident in this atlas of American change.

The boom/bust in the housing market affected not only the distribution of the population, but also the stability of the economy. Census 2010 revealed the growth of the housing market through the first half of the decade and the increase in vacant housing units in the second half. From 2000 to 2010, housing units increased by almost 14 percent. Vacant housing units increased by almost 44 percent. The momentum in housing construction did not slow as quickly as demand. The change in vacant housing exceeded the change in total housing units in one out of four counties. Foreclosures exacerbated the increasing vacancy rates and created a backlog of unsold vacant properties that will take more time to sell than it took to build.

The US vacancy rate, which is the number of vacant units relative to the total number of housing units, increased from 9 percent in 2000 to 11.4 percent in 2010. By region, the South and West posted the largest increases in housing stock, over 17 percent, and in vacant housing, more than 45 percent. However, the Midwest posted the largest increase in its vacancy rate. By state, Nevada and Georgia experienced the largest increases in vacant housing—120 percent and 83 percent. New Mexico had the smallest increase in vacant housing, just 7 percent from 2000 to 2010.

The change in households echoes the changes in the housing market. The number of households increased by almost 11 percent from 2000 to 2010, slightly faster than population change, but less than the 14 percent increase in housing units. The gap between household and housing unit change is evidenced by the increase in vacant units. The small difference between household and population growth shows the changing composition of households. Nonfamily households are increasing twice as fast as family households. Most of the growth in nonfamily households is a result of the increase in single-person households; however, the proportion of single-person households is declining. Unmarried partners are the fastest-growing household type, up by 41 percent from 2000 to 2010. Marriage rates dropped when the collapse of the housing market impacted the economy.

The domino effect of changes in the housing market increased the proportion of mortgaged homes through the first half of the decade and decreased rates of ownership in the latter half of the decade. By 2010, the share of homes with a mortgage had increased to almost 70 percent, up from 67 percent in 2000. Home ownership dropped to 65 percent, down from more than 66 percent in 2000. The change was small in the Northeast, -0.2 percentage points, and most pronounced in the South, -1.6 percentage points.

This compendium shows the population at one point in time, April 1, 2010—and the effects of change through the previous decade, 2000–2010. However, it also offers a glimpse of the future in the maps that contrast the change in the population under age 18 to the change in the population over age 18. Among adults, population aged 18 years and older, the non-Hispanic white segment continued to grow slowly, by less than 4.4 percent from 2000 to 2010. The fastest-growing segments among adults were Asian or Hispanic. Among children, the non-Hispanic white population decreased by 9.8 percent from 2000 to 2010. The fastest-growing segment among children was multiracial, with an increase of almost 46 percent in 10 years. In 2010, two out of three adults represented the majority population (non-Hispanic white). Almost one out of two children (46.5 percent) represent minority populations. The Diversity Index ranges from 37 (Midwest) to 69.5 (West) among adults; among children, the range is 53.8 (Midwest) to 81.5 (West). Although immigration declined in the latter half of the decade, the trend toward increased diversity in the American population has been set. It's reflected in the extraordinary diversity of our children.

Organization of the Atlas

This atlas is divided into four sections, each providing views of demographic data collected by the US Census Bureau. Section 1 includes snapshots of the total US population in 2000 and 2010 without reference to race or ethnicity. Section 2 is a series of maps showing the state of housing and households in America over the past decade. Section 3 maps minority populations and diversity using data from Census 2000 and Census 2010. The maps in this section allow the reader to see the racial and ethnic composition of different US regions and how those regional populations have moved and changed since 2000. Section 4 presents different views of the basic race and ethnicity data collected by the US Census. This series shows the distribution of people who identified themselves as belonging to one or more racial or ethnic groups, as well as comparisons to Census 2000 data.

Each page of the atlas features county-level detail maps for the fifty states, the District of Columbia, and Puerto Rico. Each page also includes small state-level maps for a simplified view of the population theme.

About the Maps

The maps in *Mapping Census 2010* were created using ArcGIS 10 for Desktop. Each map was drawn using a customized Albers map projection.

Most of these maps are choropleth maps. Choropleth maps display numerical data values that are divided into classes. The classes are assigned colors, which are used to shade areas on the map. The class division for the data in these maps was chosen using a combination of rounded breaks shared among maps, and national rates for each map topic. The map legends list the range of data values that each color represents.

Choropleth maps are best for showing derived values such as percent or density, and are less appropriate for representing total numbers of people. They are, however, used for totals in this atlas so that counties across maps can be easily compared.

This book focuses on map comparisons. For this reason, we have included Census 2000 data redistricted to Census 2010 county boundaries to facilitate visual comparisons and changes over the past decade.

Map Color

Maps of similar themes and maps showing data from Census 2000 and 2010 are placed on opposing pages and include color schemes deliberately chosen for optimal visual comparison.

Eight color schemes are used in this book. In sections 1, 2, and 3 several combinations of these schemes are used to represent the data. High values are shown in dark blue, purple, and green. Low or negative values are shown in pink and orange hues. The maps in section 4 display three specific and consistent color schemes. Purple is used for absolute population, dark blue to light green are used for percentages of the population, and a diverging dual color scheme is used to show changes in percentages of the population. Dark purple indicates increases in populations within that county, and dark orange indicates the greatest decreases in population.

County boundaries are shown in a slightly darker hue than the value within the polygon with the intent of placing more visual emphasis on patterns and clusters. ArcGIS for Desktop assigned different segments of an area to be displayed on top of others based on symbol levels.

Data ranges for colors change when breaks are adjusted to include US overall rates specific to the group mapped.

About the Data

This atlas maps the range of information that the Census Bureau collected in Census 2010. All datasets were rounded to one decimal. Large numbers in the maximum category on the Number of People map legends are rounded up to the next highest number with four significant digits.

The percent of US population for a particular group is calculated using populations for the United States as a whole; for example, the total population less than 18 years is divided by the total US population. Note that these summary numbers are not averages of state or county percentages. The calculations of the US summary numbers are based on data from all states and the District of Columbia, but do not include Puerto Rico.

In some cases, an overlay was used to mask counties containing very small populations of the group mapped. This was done because small populations often produce extreme percentage changes that distract from the more reliable pattern seen over areas with greater numbers of people. A threshold of 100 people indicating that race or category from the 2010 Census was used to create the mask. So, for example, the percent change in populations of Hispanic or Latino origin is not shown in counties where there are fewer than 100 Hispanics in 2010.

Changes from 2000 to 2010

The maps showing 2000 to 2010 percent change compare data from Census 2000 to Census 2010 counts. Census 2000 data on race is directly comparable to Census 2010 data, which offers a real measure of population change in individual race groups and the multiracial group from 2000 to 2010. Persons of Hispanic origin may be of any race.

One change for Census 2010 was the inclusion of mortgage status as a complete-count question. In 2000, mortgage status was tabulated as a sample item. Another change in 2010 is the report of multigenerational households, which include three or more parent-child generations. Census 2010 also did not enumerate foster children separately. Therefore, the map displaying households with children actually represents households with population under 18 years of age. This tabulation was selected to include foster children, but it may include persons less than 18 years who are not related or foster children.

Previously published Census 2000 data was reaggregated to Census 2010 geography. Changes in the areas for which data are tabulated and reported are critical to the analysis of trends. Larger political areas like counties change less often than cities or townships, but boundary revisions are common among counties in Census 2010. The only way to compare the data from census to census is to define a correspondence between the geographic areas in each.

Esri defined the 2000–2010 correspondence by using different sources of data from the Census Bureau. The first step was to create a spatial join with the 2000 blocks and the 2010 boundaries for geographic areas from TIGER 2010. The technique aligns the internal point, expressed as latitude-longitude coordinates, of 2000 blocks with 2010 boundaries. Next, extensive edits of the assignment of 2000 blocks to 2010 areas were necessary to ensure comparability to the Census Bureau's internal process of allocating addresses to specific areas. Assignments were checked against the Census Bureau's 2000/2010 tabulation block correspondence.

The use of more than eight million census blocks to depict the location of the population raises the development of geographic correspondence from an art to a science. By overlaying 2010 boundaries on the landscape of blocks, it is possible to treat the distribution of the population as continuous, rather than discrete and unique to each small statistical area. This resolution clarifies the correspondence between 2000 and 2010 geography and provides a consistent 2000 database in 2010 geography.

- Riley Peake, Cartographer, Esri
- Lynn Wombold, Chief Demographer, Esri

Section

1

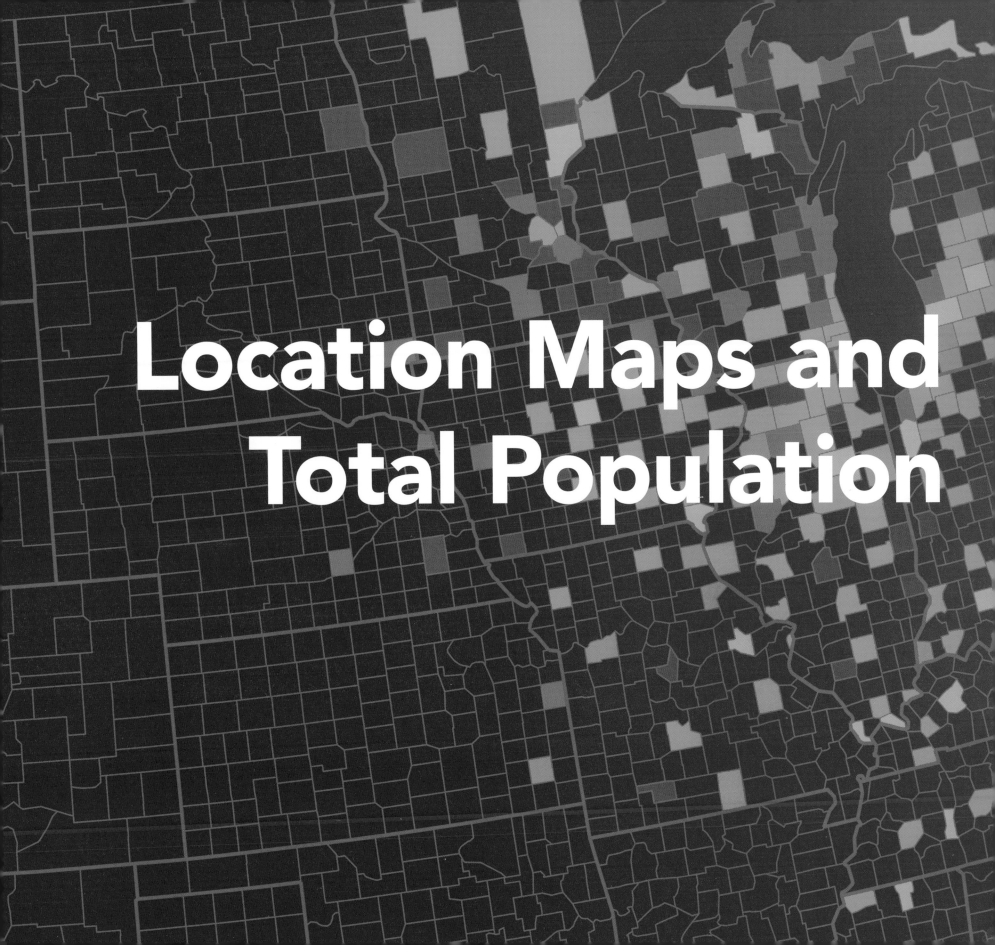

Location Maps and Total Population

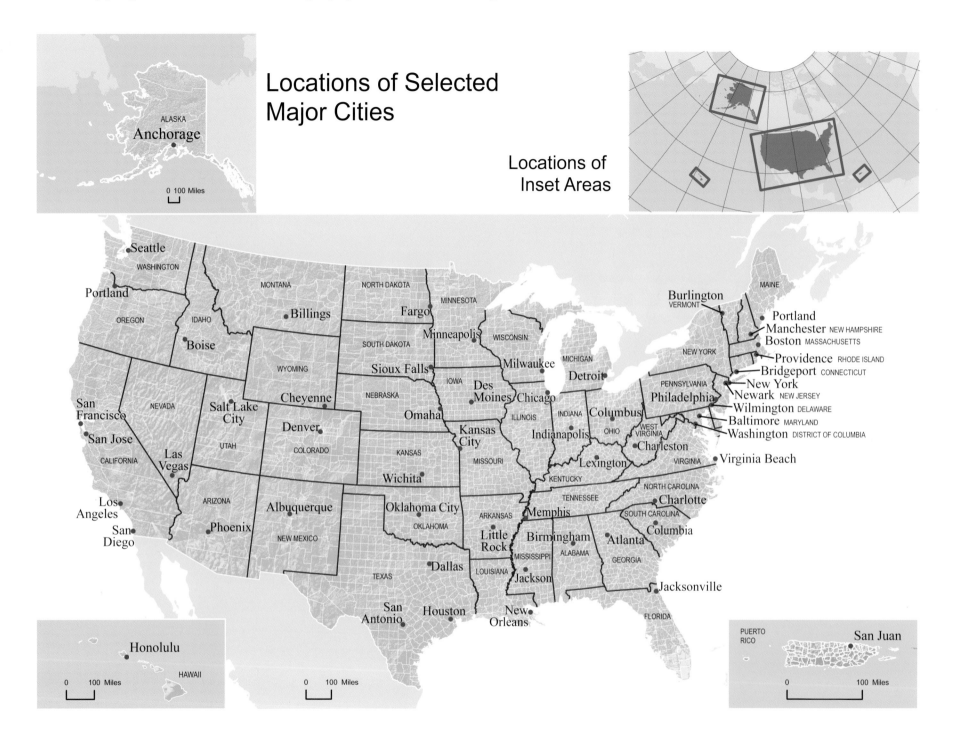

Locations of Selected Major Cities

Locations of Inset Areas

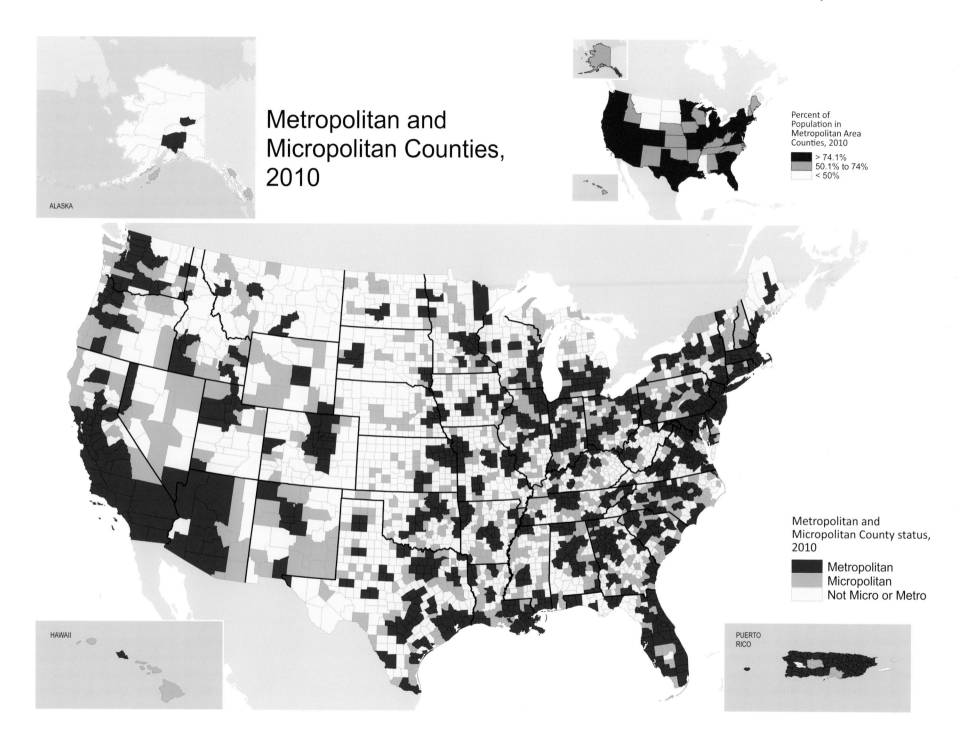

Metropolitan and Micropolitan Counties, 2010

ALASKA

Percent of Population in Metropolitan Area Counties, 2010

- > 74.1%
- 50.1% to 74%
- < 50%

Metropolitan and Micropolitan County status, 2010

- Metropolitan
- Micropolitan
- Not Micro or Metro

HAWAII

PUERTO RICO

Metropolitan and micropolitan status are determined by the Office of Management and Budget (OMB) for Core Based Statistical Areas (CBSA). Current designations are based on 2009 OMB classifications. CBSAs consist of counties and equivalent entities. CBSAs are categorized based on the population of the largest urban area within the CBSA. Categories of CBSAs are Metropolitan Statistical Areas, based on urbanized areas of 50,000 or more population, and Micropolitan Statistical Areas, based on urban clusters of population at least 10,000 but less than 50,000.

Total Population, 2000

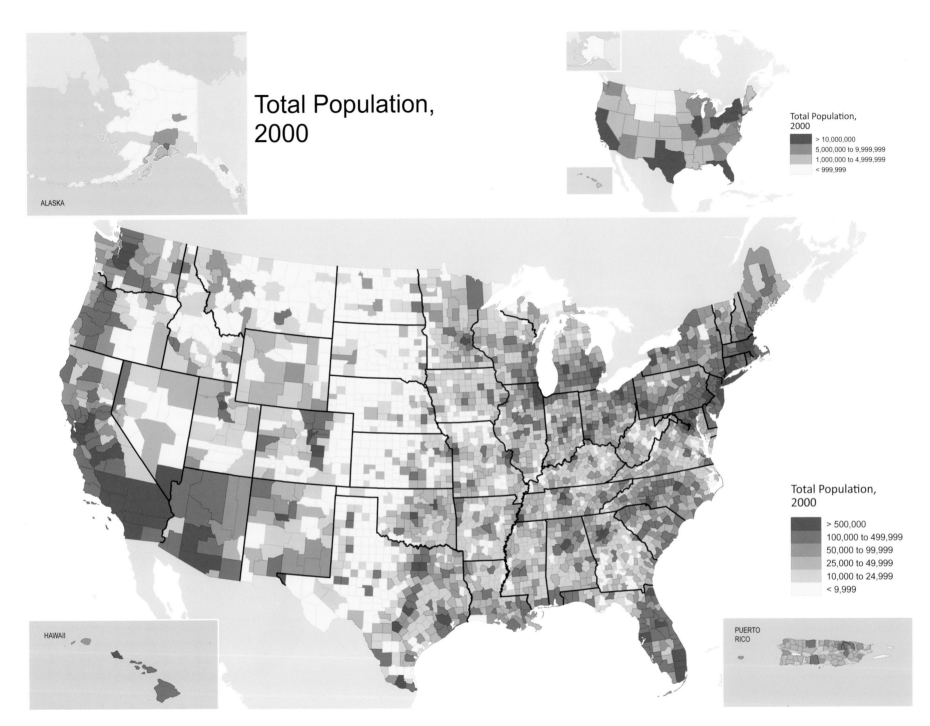

Total Population, 2000

- > 10,000,000
- 5,000,000 to 9,999,999
- 1,000,000 to 4,999,999
- < 999,999

ALASKA

Total Population, 2000

- > 500,000
- 100,000 to 499,999
- 50,000 to 99,999
- 25,000 to 49,999
- 10,000 to 24,999
- < 9,999

HAWAII

PUERTO RICO

Total Population, 2010

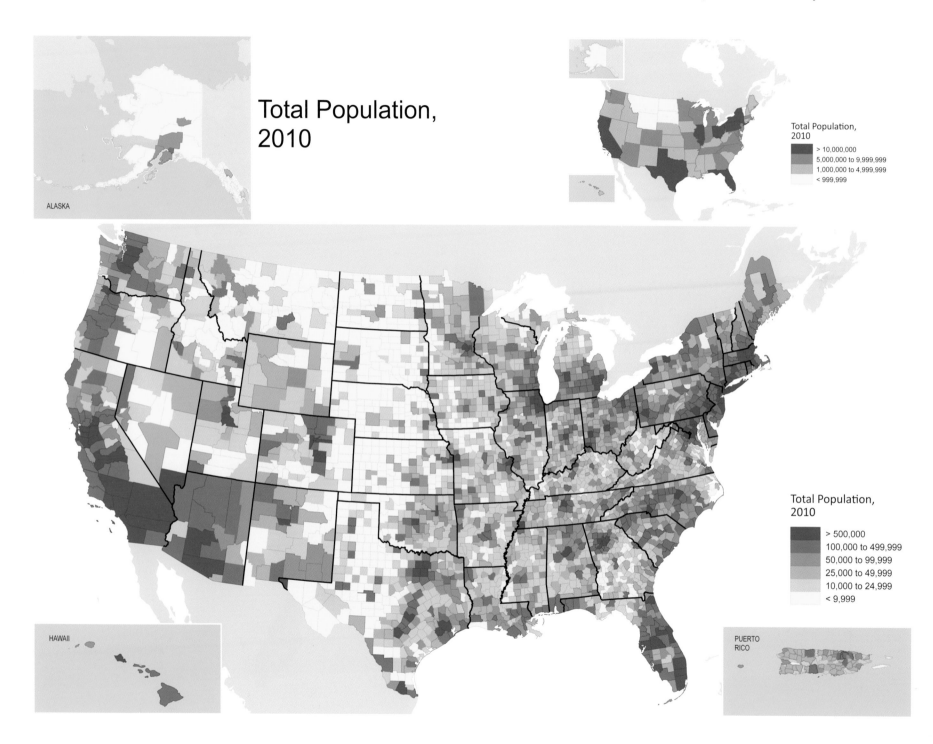

Total Population, 2010
- \> 10,000,000
- 5,000,000 to 9,999,999
- 1,000,000 to 4,999,999
- < 999,999

Total Population, 2010
- \> 500,000
- 100,000 to 499,999
- 50,000 to 99,999
- 25,000 to 49,999
- 10,000 to 24,999
- < 9,999

ALASKA

HAWAII

PUERTO RICO

Population Density, 2000

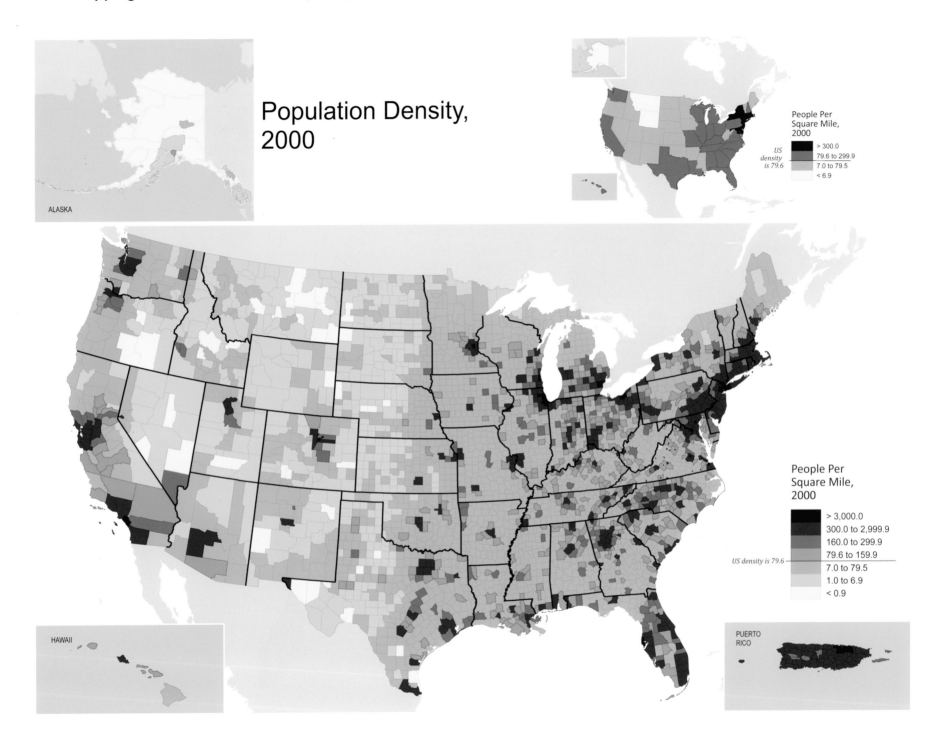

ALASKA

People Per
Square Mile,
2000

US
density
is 79.6

> 300.0
79.6 to 299.9
7.0 to 79.5
< 6.9

People Per
Square Mile,
2000

> 3,000.0
300.0 to 2,999.9
160.0 to 299.9
79.6 to 159.9

US density is 79.6

7.0 to 79.5
1.0 to 6.9
< 0.9

HAWAII

PUERTO
RICO

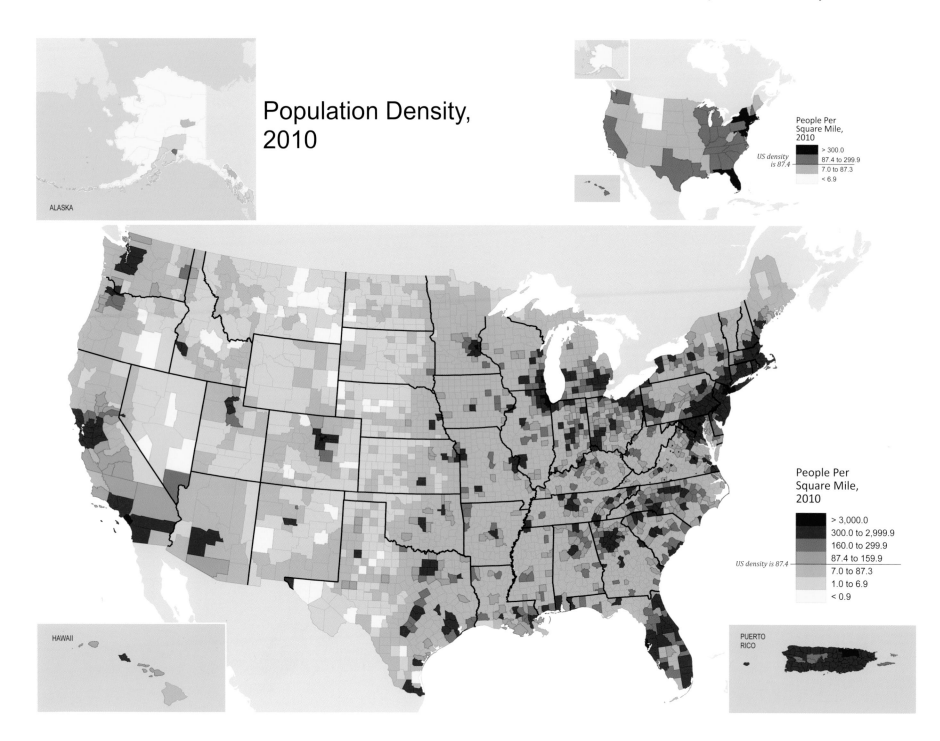

Population Density, 2010

People Per Square Mile, 2010

US density is 87.4

- > 300.0
- 87.4 to 299.9
- 7.0 to 87.3
- < 6.9

ALASKA

People Per Square Mile, 2010

- > 3,000.0
- 300.0 to 2,999.9
- 160.0 to 299.9
- 87.4 to 159.9

US density is 87.4

- 7.0 to 87.3
- 1.0 to 6.9
- < 0.9

HAWAII

PUERTO RICO

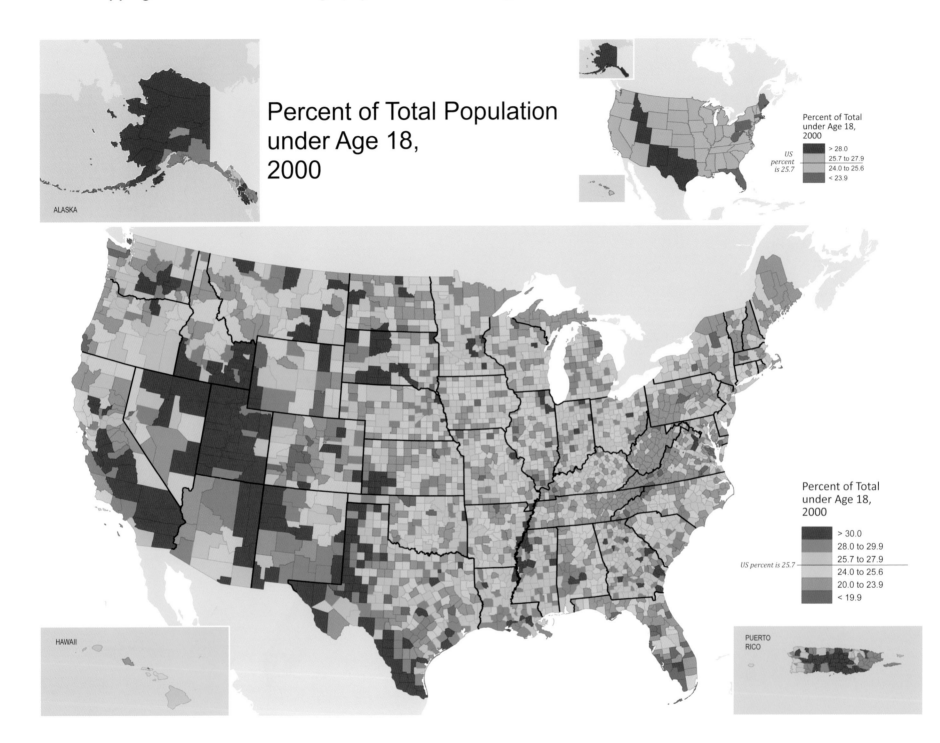

Percent of Total Population under Age 18, 2000

ALASKA

Percent of Total under Age 18, 2000

US percent is 25.7

> 28.0
25.7 to 27.9
24.0 to 25.6
< 23.9

Percent of Total under Age 18, 2000

> 30.0
28.0 to 29.9
25.7 to 27.9
US percent is 25.7
24.0 to 25.6
20.0 to 23.9
< 19.9

HAWAII

PUERTO RICO

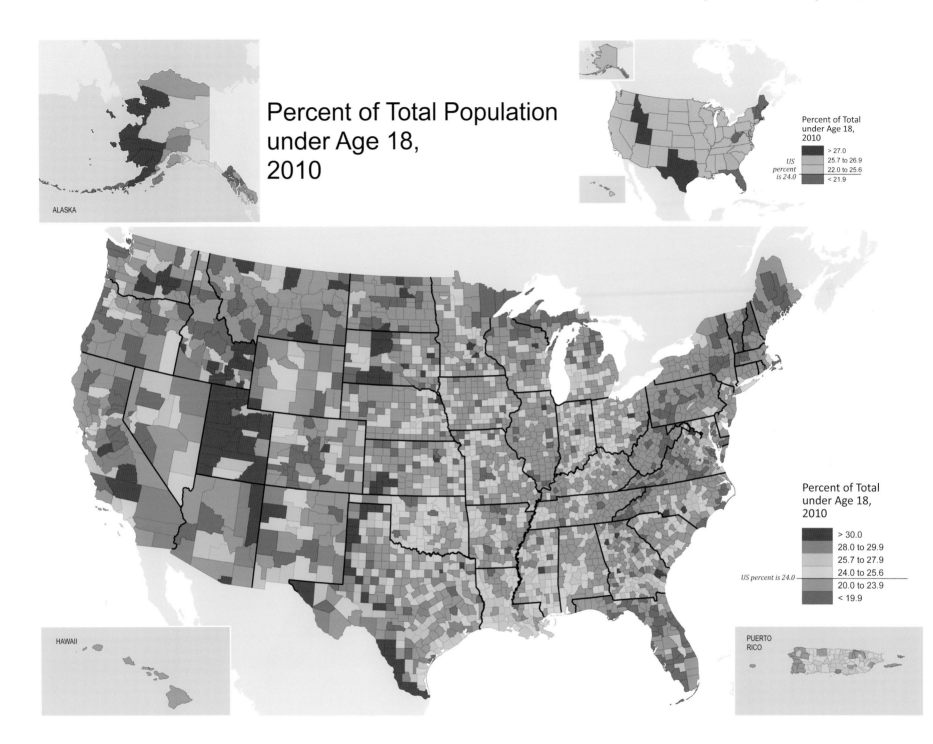

Percent of Total Population
under Age 18,
2010

Percent of Total
under Age 18,
2010

US
percent
is 24.0

> 27.0
25.7 to 26.9
22.0 to 25.6
< 21.9

ALASKA

Percent of Total
under Age 18,
2010

> 30.0
28.0 to 29.9
25.7 to 27.9
24.0 to 25.6
20.0 to 23.9
< 19.9

US percent is 24.0

HAWAII

PUERTO
RICO

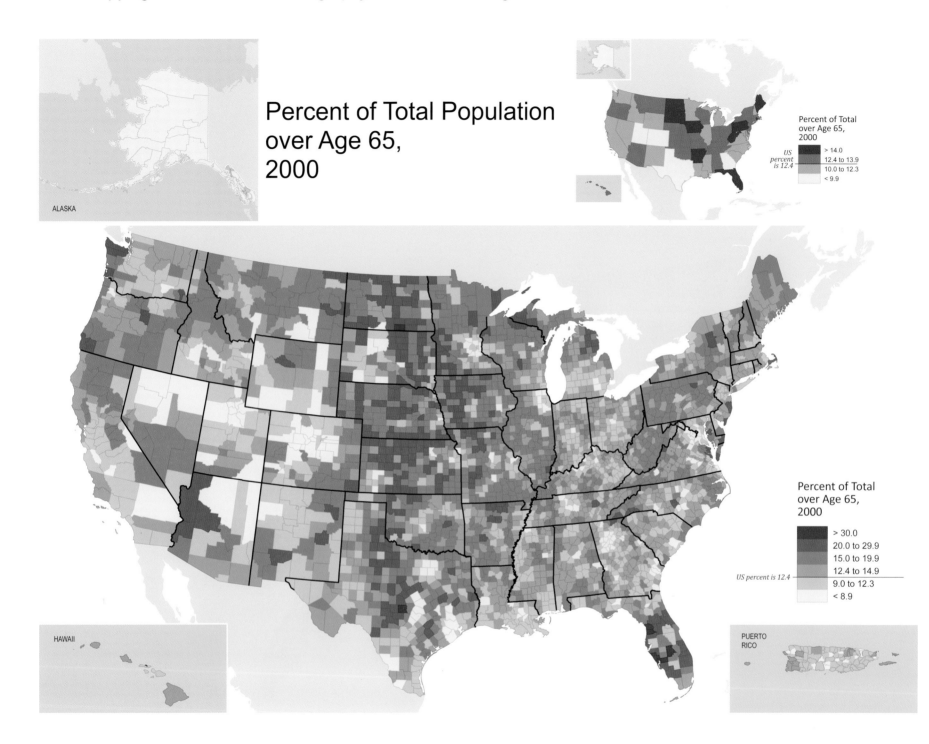

Percent of Total Population
over Age 65,
2000

Percent of Total
over Age 65,
2000

US
percent
is 12.4

> 14.0
12.4 to 13.9
10.0 to 12.3
< 9.9

ALASKA

Percent of Total
over Age 65,
2000

> 30.0
20.0 to 29.9
15.0 to 19.9
12.4 to 14.9
9.0 to 12.3
< 8.9

US percent is 12.4

HAWAII

PUERTO
RICO

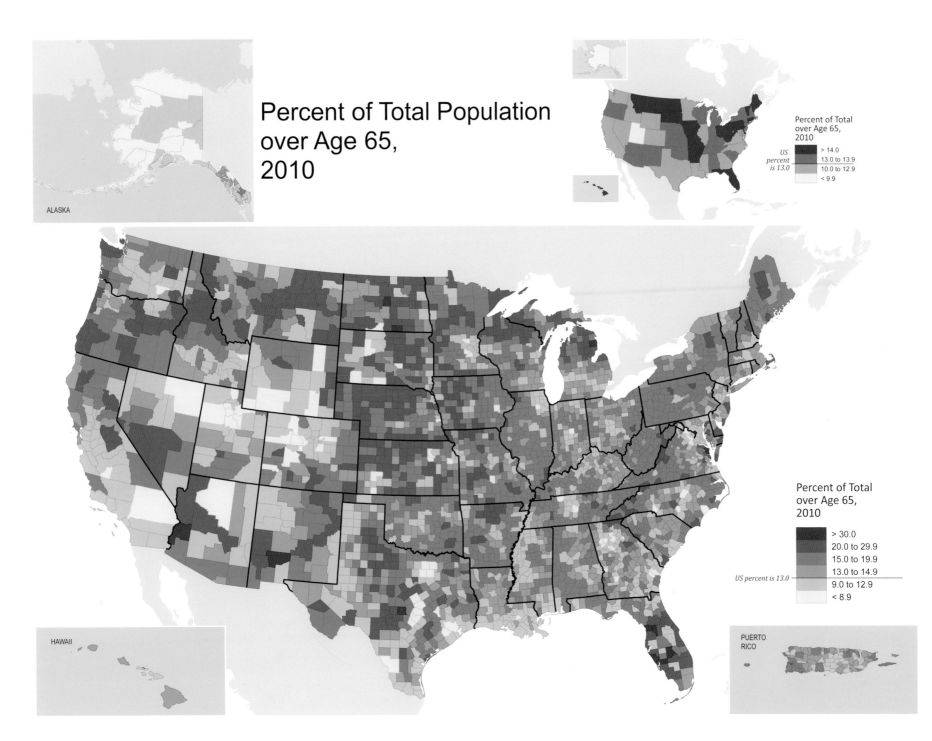

Percent of Total Population
over Age 65,
2010

ALASKA

Percent of Total
over Age 65,
2010

US percent is 13.0

> 14.0
13.0 to 13.9
10.0 to 12.9
< 9.9

Percent of Total
over Age 65,
2010

> 30.0
20.0 to 29.9
15.0 to 19.9
13.0 to 14.9
9.0 to 12.9
< 8.9

US percent is 13.0

HAWAII

PUERTO
RICO

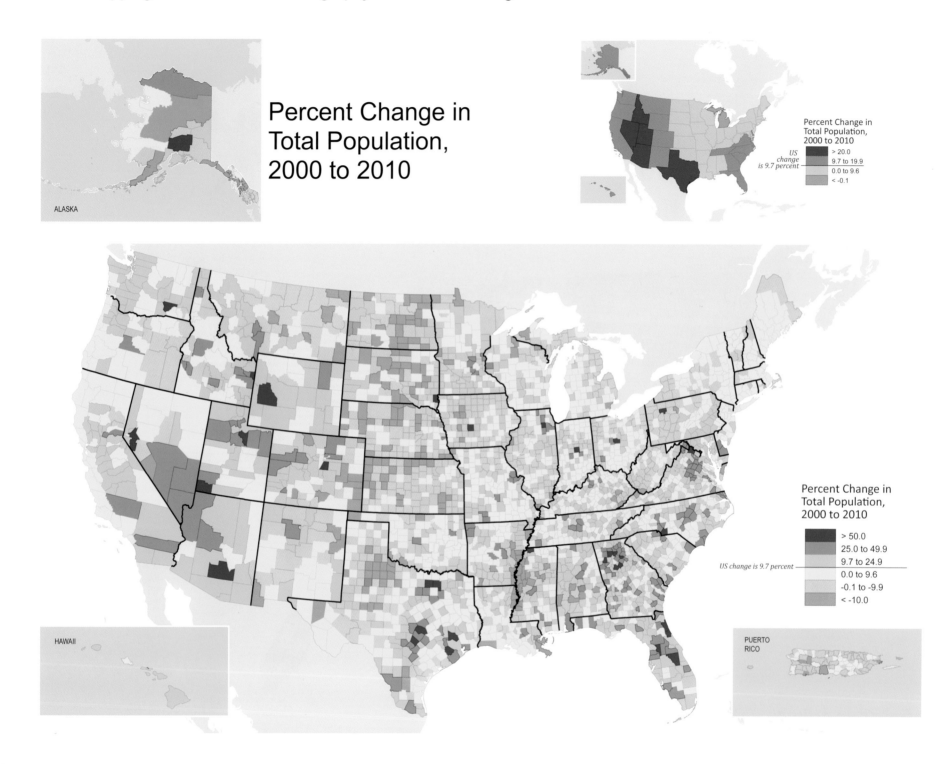

ALASKA

Percent Change in
Total Population,
2000 to 2010

Percent Change in
Total Population,
2000 to 2010

US
change
is 9.7 percent
> 20.0
9.7 to 19.9
0.0 to 9.6
< -0.1

Percent Change in
Total Population,
2000 to 2010

> 50.0
25.0 to 49.9
9.7 to 24.9
US change is 9.7 percent
0.0 to 9.6
-0.1 to -9.9
< -10.0

HAWAII

PUERTO
RICO

Median Age, 2010

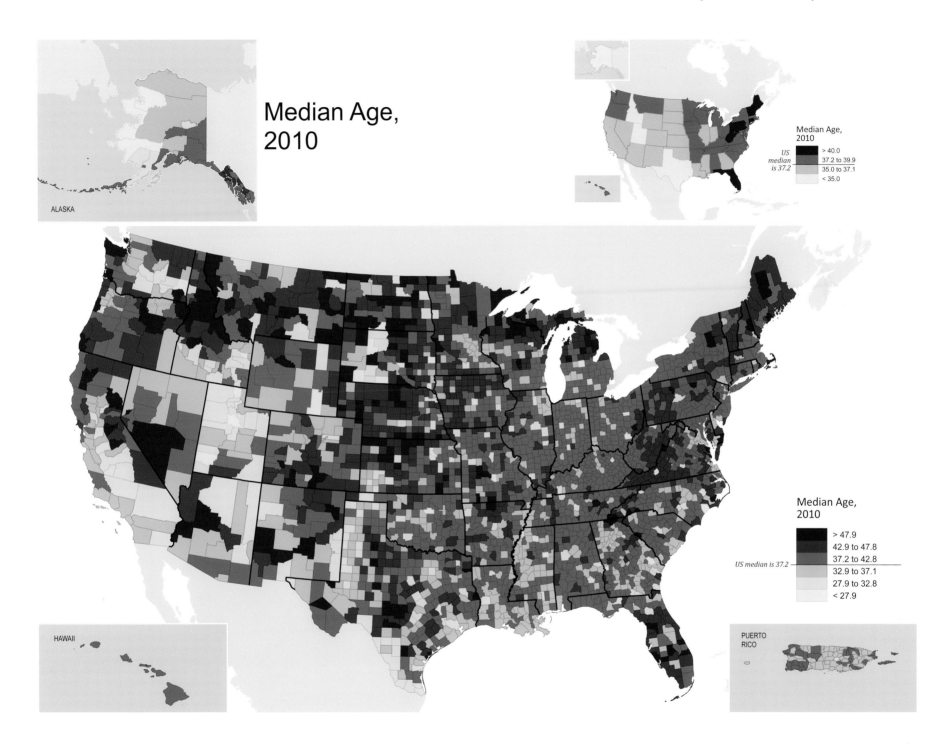

ALASKA

Median Age, 2010

US median is 37.2

- > 40.0
- 37.2 to 39.9
- 35.0 to 37.1
- < 35.0

Median Age, 2010

- > 47.9
- 42.9 to 47.8
- 37.2 to 42.8
- US median is 37.2 ——
- 32.9 to 37.1
- 27.9 to 32.8
- < 27.9

HAWAII

PUERTO RICO

Percent Change in Population under Age 18, 2000 to 2010

ALASKA

Percent Change under Age 18, 2000 to 2010

US change is 2.6 percent

- > 10.0
- 2.6 to 9.9
- 0.0 to 2.5
- < -0.1

Percent Change under Age 18, 2000 to 2010

US change is 2.6 percent

- > 50.0
- 25.0 to 49.9
- 2.6 to 24.9
- 0.0 to 2.5
- -0.1 to -9.9
- < -10.0

HAWAII

PUERTO RICO

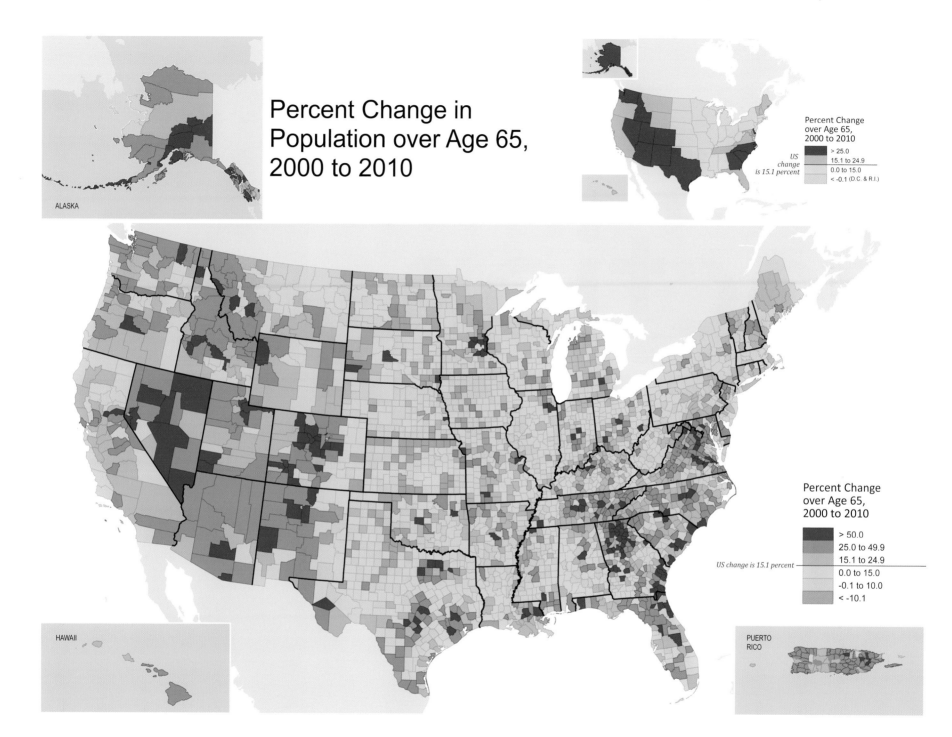

Percent Change in
Population over Age 65,
2000 to 2010

ALASKA

Percent Change
over Age 65,
2000 to 2010

US
change
is 15.1 percent

> 25.0
15.1 to 24.9
0.0 to 15.0
< -0.1 (D.C. & R.I.)

Percent Change
over Age 65,
2000 to 2010

US change is 15.1 percent

> 50.0
25.0 to 49.9
15.1 to 24.9
0.0 to 15.0
-0.1 to 10.0
< -10.1

HAWAII

PUERTO
RICO

Section
2

Households

Percent of Family Households, 2000

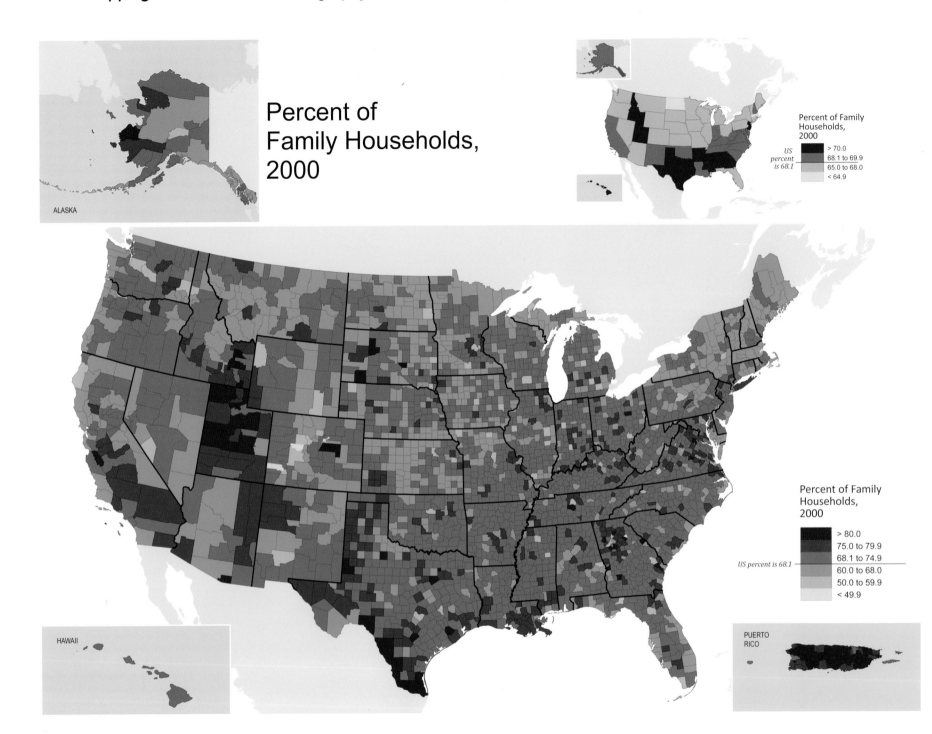

Percent of Family Households, 2000

	> 70.0
US percent is 68.1	68.1 to 69.9
	65.0 to 68.0
	< 64.9

ALASKA

Percent of Family Households, 2000

	> 80.0
	75.0 to 79.9
	68.1 to 74.9
US percent is 68.1	60.0 to 68.0
	50.0 to 59.9
	< 49.9

HAWAII

PUERTO RICO

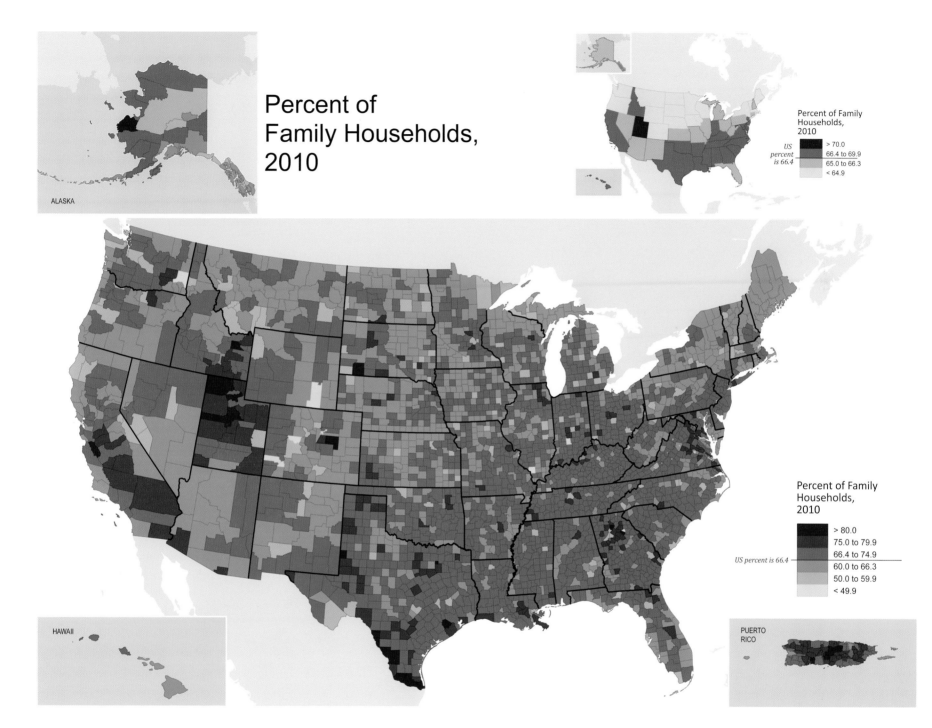

Percent of
Family Households,
2010

ALASKA

Percent of Family
Households,
2010

US percent is 66.4

▉	> 70.0
▉	66.4 to 69.9
▉	65.0 to 66.3
▉	< 64.9

HAWAII

PUERTO RICO

Percent of Family
Households,
2010

US percent is 66.4

▉	> 80.0
▉	75.0 to 79.9
▉	66.4 to 74.9
▉	60.0 to 66.3
▉	50.0 to 59.9
▉	< 49.9

 A family household represents a household where the householder is related to at least one member of the household by birth, marriage, or adoption.

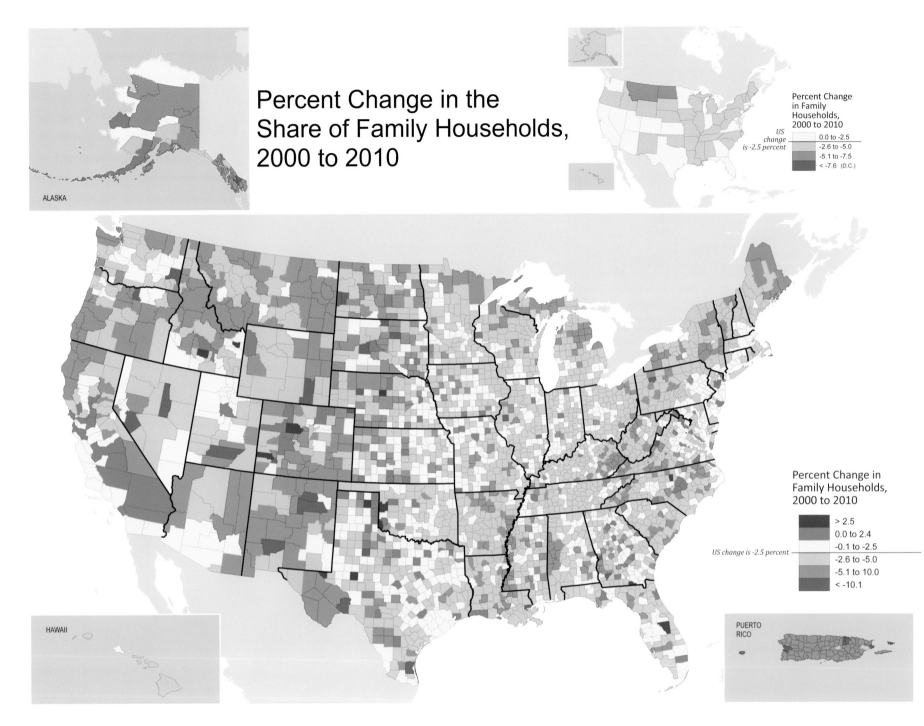

Percent Change in the Share of Family Households, 2000 to 2010

ALASKA

Percent Change in Family Households, 2000 to 2010

US change is -2.5 percent

	0.0 to -2.5
	-2.6 to -5.0
	-5.1 to -7.5
	< -7.6 (D.C.)

Percent Change in Family Households, 2000 to 2010

	> 2.5
	0.0 to 2.4
	-0.1 to -2.5
	-2.6 to -5.0
	-5.1 to 10.0
	< -10.1

US change is -2.5 percent

HAWAII

PUERTO RICO

A family household represents a household where the householder is related to at least one member of the household by birth, marriage, or adoption.

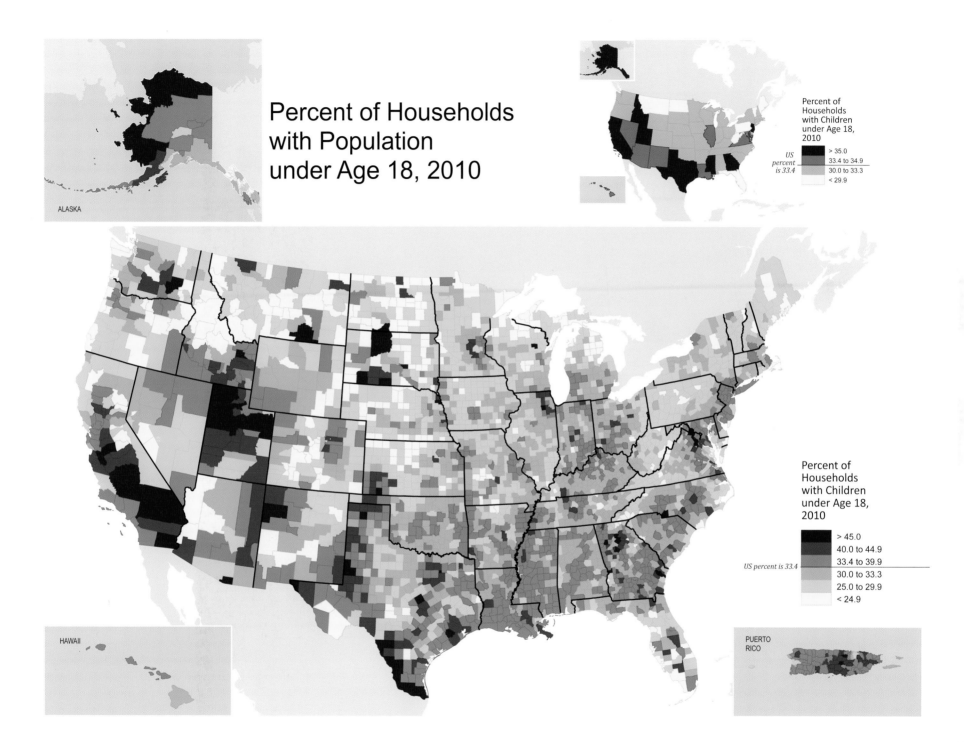

Percent of Households
with Population
under Age 18, 2010

Percent of
Households
with Children
under Age 18,
2010

US
percent
is 33.4

> 35.0
33.4 to 34.9
30.0 to 33.3
< 29.9

ALASKA

Percent of
Households
with Children
under Age 18,
2010

US percent is 33.4

> 45.0
40.0 to 44.9
33.4 to 39.9
30.0 to 33.3
25.0 to 29.9
< 24.9

HAWAII

PUERTO
RICO

Percent of Multigenerational Households, 2010

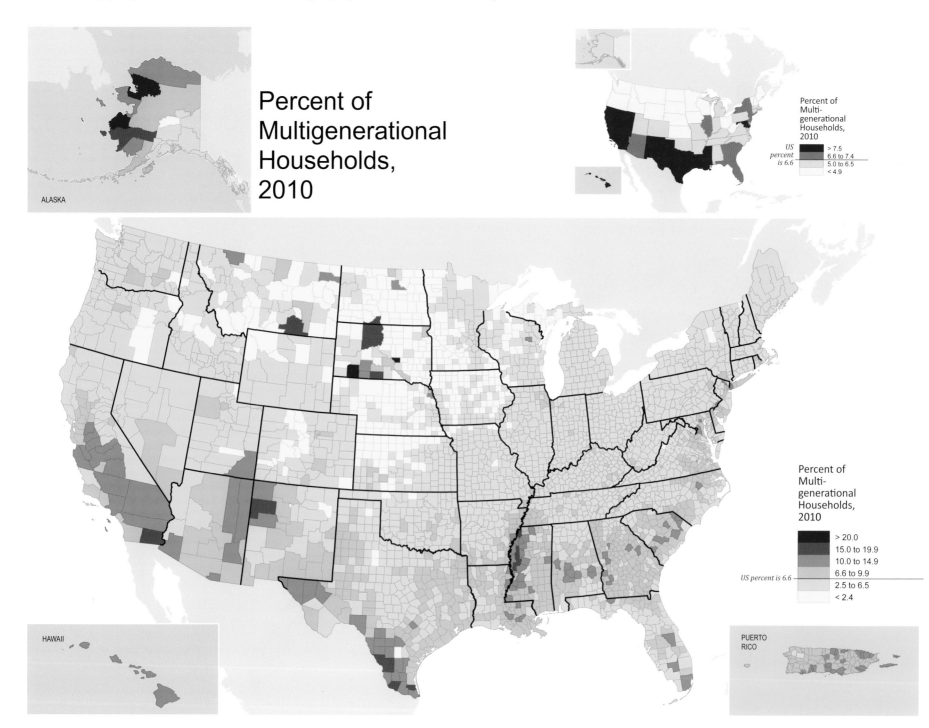

ALASKA

Percent of Multi-generational Households, 2010

US percent is 6.6

- > 7.5
- 6.6 to 7.4
- 5.0 to 6.5
- < 4.9

Percent of Multi-generational Households, 2010

- > 20.0
- 15.0 to 19.9
- 10.0 to 14.9
- 6.6 to 9.9
- 2.5 to 6.5
- < 2.4

US percent is 6.6

HAWAII

PUERTO RICO

Multigenerational households contain three or more parent-child generations. Parent-child generations can be either biological, by adoption, or through marriage, i.e., stepfamily.

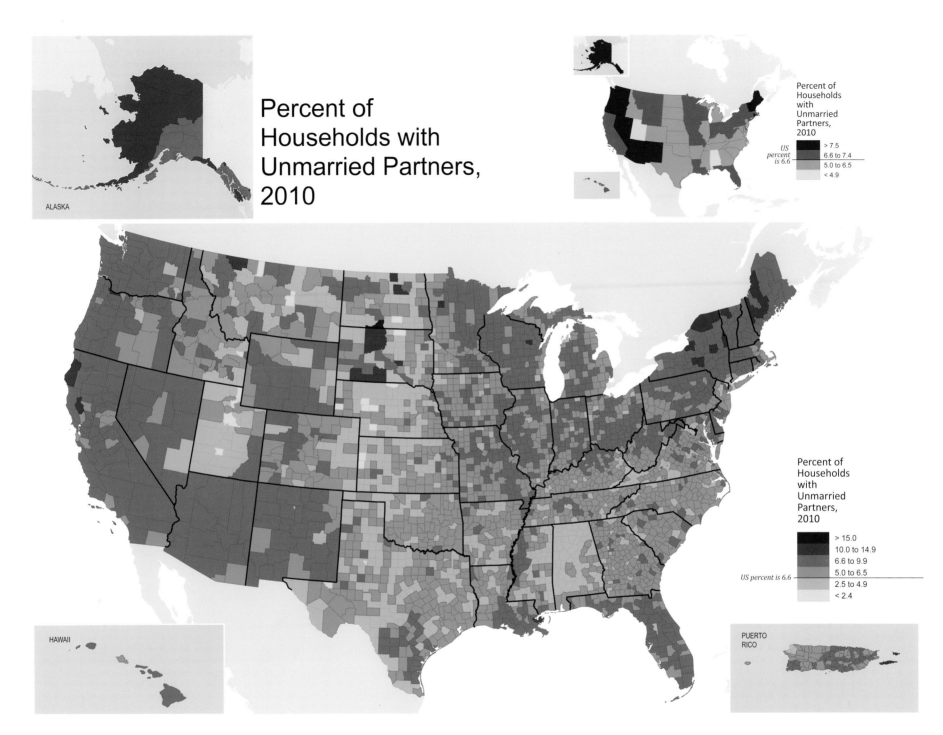

Percent of Households with Unmarried Partners, 2010

ALASKA

Percent of Households with Unmarried Partners, 2010

US percent is 6.6

- > 7.5
- 6.6 to 7.4
- 5.0 to 6.5
- < 4.9

Percent of Households with Unmarried Partners, 2010

- > 15.0
- 10.0 to 14.9
- 6.6 to 9.9
- 5.0 to 6.5
- 2.5 to 4.9
- < 2.4

US percent is 6.6

HAWAII

PUERTO RICO

Unmarried partner households include a householder and his or her unmarried partner. The unmarried partner in these households shares living quarters and has a close personal relationship with the householder. The unmarried partner in these households can be the same sex or the opposite sex as the householder. The US Census Bureau includes same-sex married-couple households in this category.

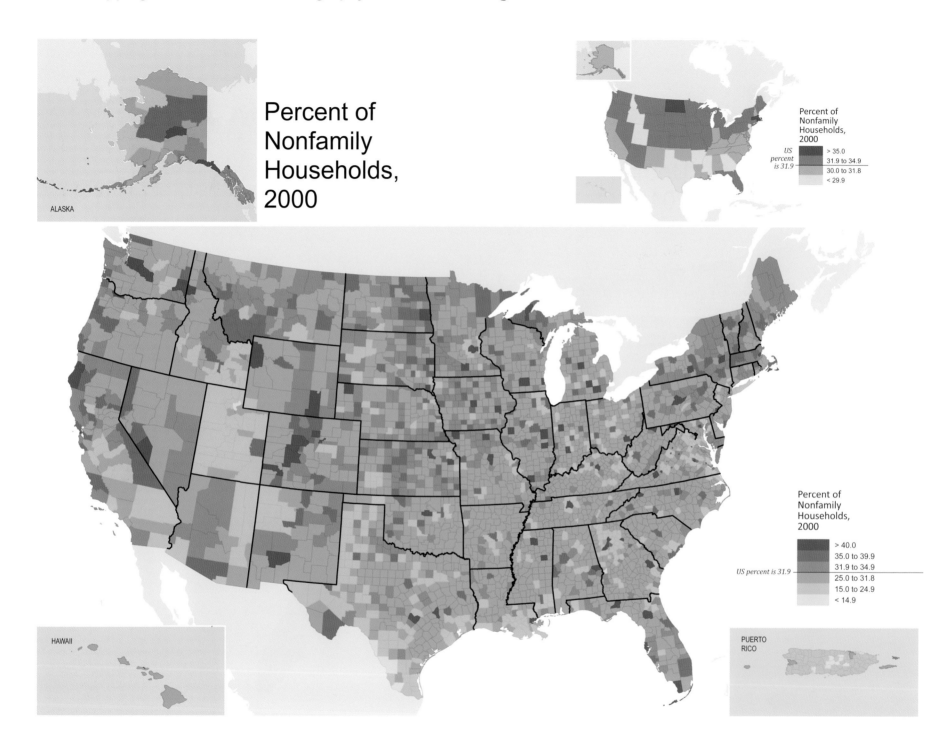

Percent of
Nonfamily
Households,
2000

ALASKA

Percent of
Nonfamily
Households,
2000

US
percent
is 31.9

> 35.0
31.9 to 34.9
30.0 to 31.8
< 29.9

Percent of
Nonfamily
Households,
2000

US percent is 31.9

> 40.0
35.0 to 39.9
31.9 to 34.9
25.0 to 31.8
15.0 to 24.9
< 14.9

HAWAII

PUERTO
RICO

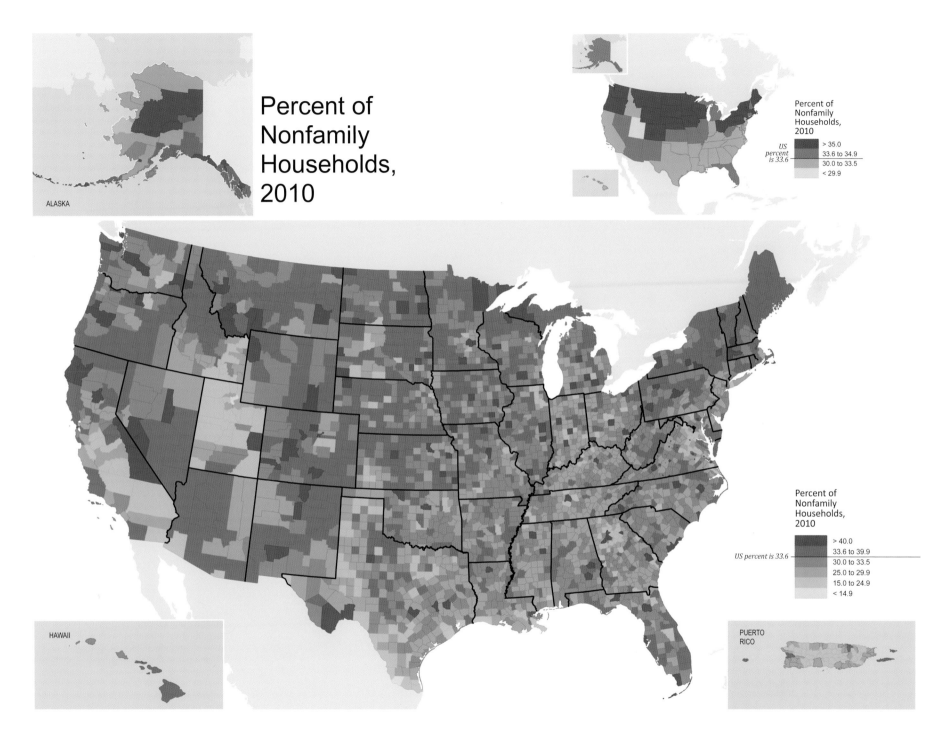

ALASKA

Percent of Nonfamily Households, 2010

Percent of
Nonfamily
Households,
2010

US
percent
is 33.6

> 35.0
33.6 to 34.9
30.0 to 33.5
< 29.9

Percent of
Nonfamily
Households,
2010

US percent is 33.6

> 40.0
33.6 to 39.9
30.0 to 33.5
25.0 to 29.9
15.0 to 24.9
< 14.9

HAWAII

PUERTO
RICO

Nonfamily households are households where the householder lives alone or lives with unrelated persons. Households with same-sex couples that are not related to the householder are also considered nonfamily households.

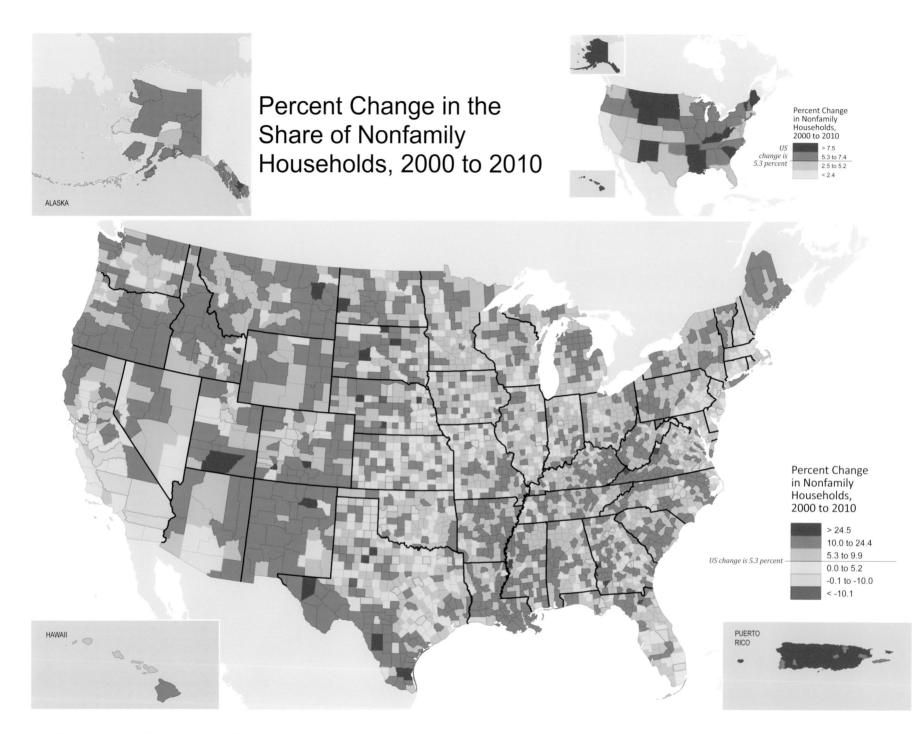

Percent Change in the Share of Nonfamily Households, 2000 to 2010

ALASKA

Percent Change in Nonfamily Households, 2000 to 2010

US change is 5.3 percent

- > 7.5
- 5.3 to 7.4
- 2.5 to 5.2
- < 2.4

HAWAII

PUERTO RICO

Percent Change in Nonfamily Households, 2000 to 2010

- > 24.5
- 10.0 to 24.4
- 5.3 to 9.9

US change is 5.3 percent

- 0.0 to 5.2
- -0.1 to -10.0
- < -10.1

Nonfamily households are households where the householder lives alone or lives with unrelated persons. Households with same-sex couples that are not related to the householder are also considered nonfamily households.

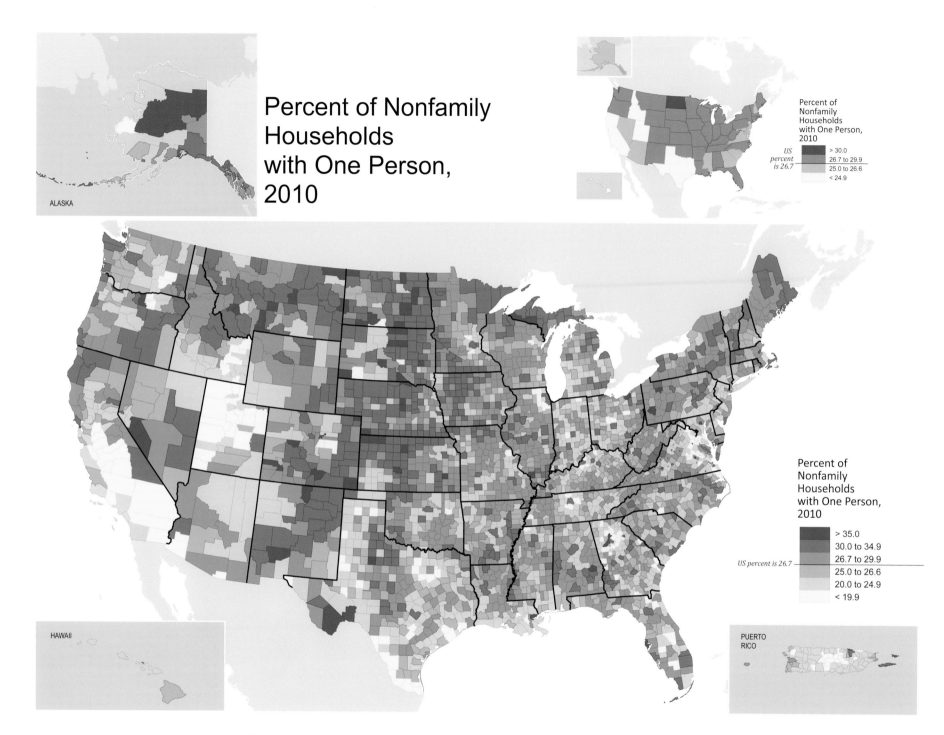

Percent of Nonfamily Households with One Person, 2010

ALASKA

Percent of Nonfamily Households with One Person, 2010

US percent is 26.7

> 30.0
26.7 to 29.9
25.0 to 26.6
< 24.9

Percent of Nonfamily Households with One Person, 2010

> 35.0
30.0 to 34.9
26.7 to 29.9
US percent is 26.7 —
25.0 to 26.6
20.0 to 24.9
< 19.9

HAWAII

PUERTO RICO

▲   Nonfamily households with one person are households where the householder lives alone.

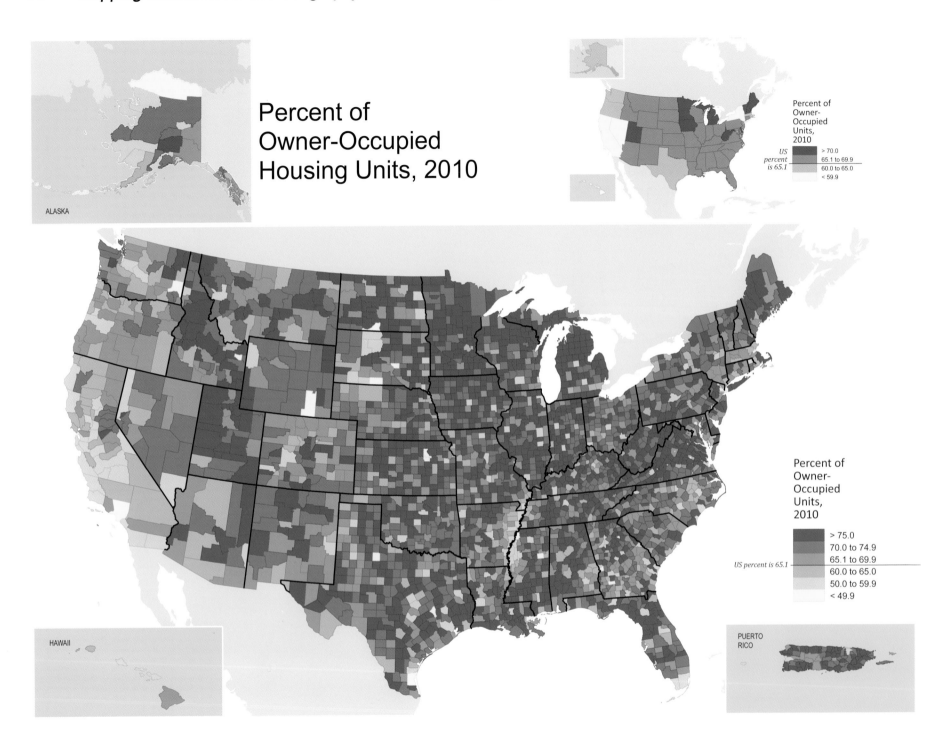

Percent of Owner-Occupied Housing Units, 2010

ALASKA

Percent of
Owner-
Occupied
Units,
2010

US percent is 65.1	> 70.0
	65.1 to 69.9
	60.0 to 65.0
	< 59.9

Percent of
Owner-
Occupied
Units,
2010

	> 75.0
	70.0 to 74.9
	65.1 to 69.9
US percent is 65.1	60.0 to 65.0
	50.0 to 59.9
	< 49.9

HAWAII

PUERTO RICO

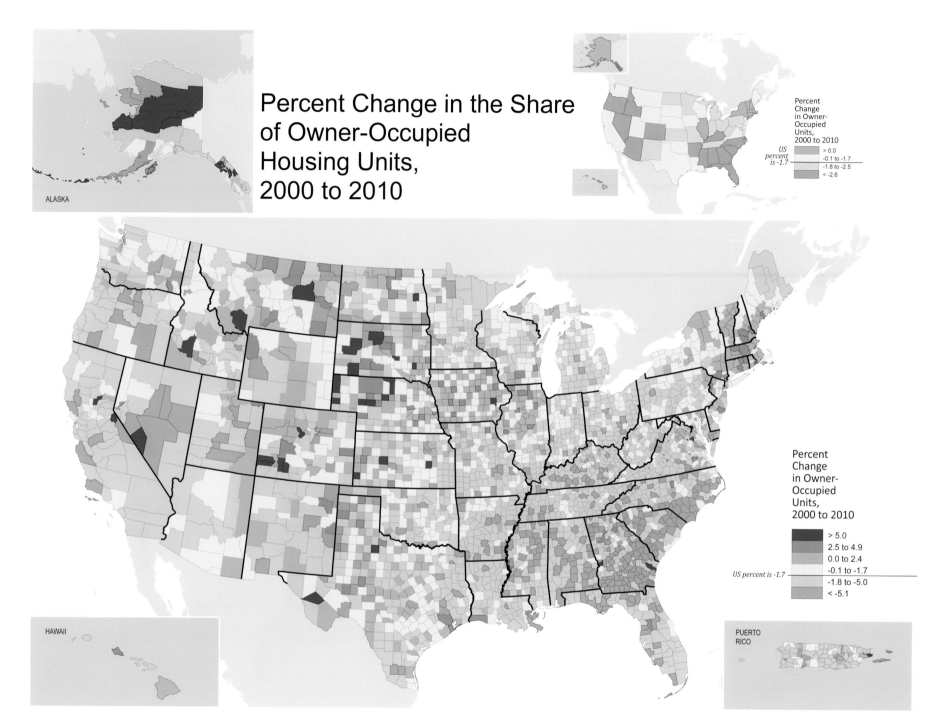

Percent Change in the Share of Owner-Occupied Housing Units, 2000 to 2010

ALASKA

Percent
Change
in Owner-
Occupied
Units,
2000 to 2010

US
percent
is -1.7

> 0.0
-0.1 to -1.7
-1.8 to -2.5
< -2.6

Percent
Change
in Owner-
Occupied
Units,
2000 to 2010

> 5.0
2.5 to 4.9
0.0 to 2.4
-0.1 to -1.7
-1.8 to -5.0
< -5.1

US percent is -1.7

HAWAII

PUERTO
RICO

◀ ▲   Owner-occupied housing units are housing units that are owned outright or mortgaged and are lived in by the owner or co-owner.

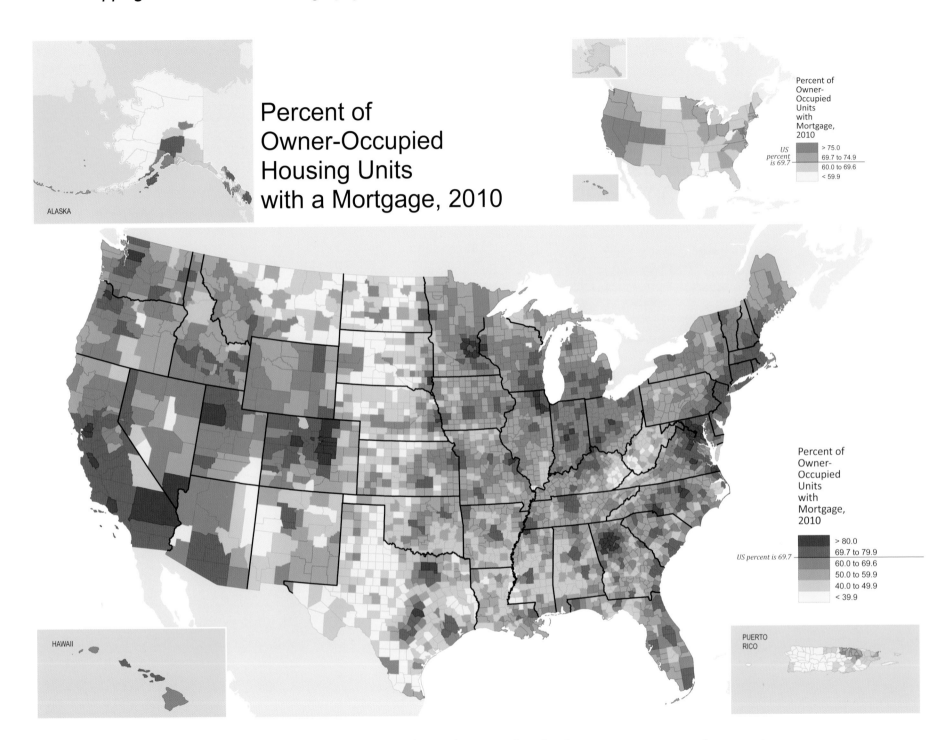

Percent of Owner-Occupied Housing Units with a Mortgage, 2010

ALASKA

Percent of Owner-Occupied Units with Mortgage, 2010

US percent is 69.7

	> 75.0
	69.7 to 74.9
	60.0 to 69.6
	< 59.9

Percent of Owner-Occupied Units with Mortgage, 2010

US percent is 69.7

	> 80.0
	69.7 to 79.9
	60.0 to 69.6
	50.0 to 59.9
	40.0 to 49.9
	< 39.9

HAWAII

PUERTO RICO

Owner-occupied housing units with a mortgage are housing units that are being purchased with a mortgage or some other debt agreement and are lived in by the owner or co-owner.

Housing

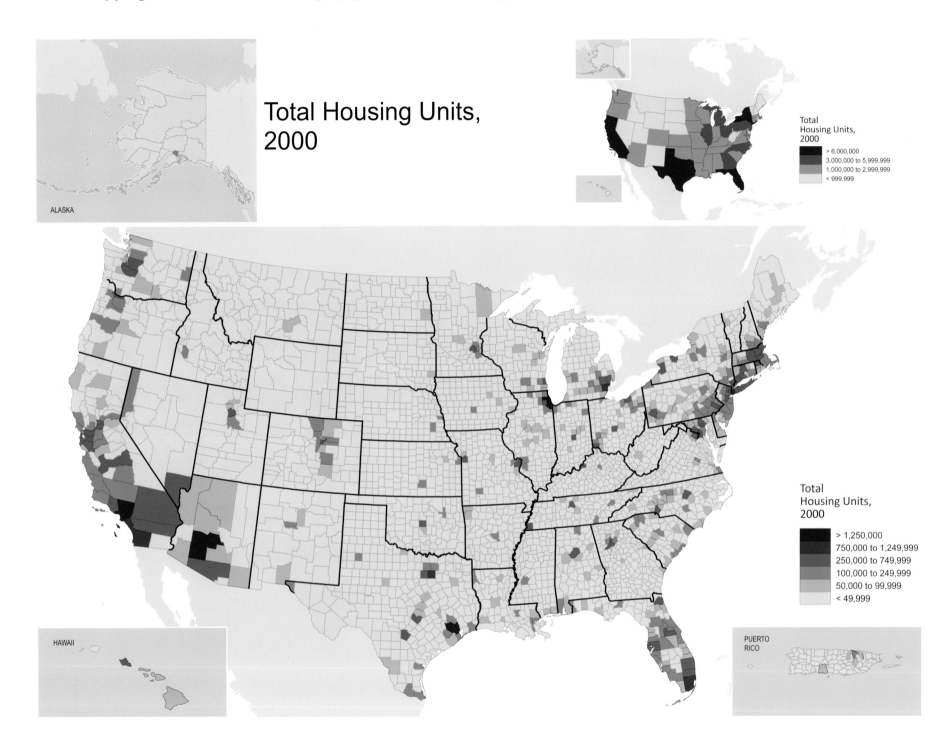

Total Housing Units,
2000

Total
Housing Units,
2000

> 6,000,000
3,000,000 to 5,999,999
1,000,000 to 2,999,999
< 999,999

ALASKA

Total
Housing Units,
2000

> 1,250,000
750,000 to 1,249,999
250,000 to 749,999
100,000 to 249,999
50,000 to 99,999
< 49,999

HAWAII

PUERTO
RICO

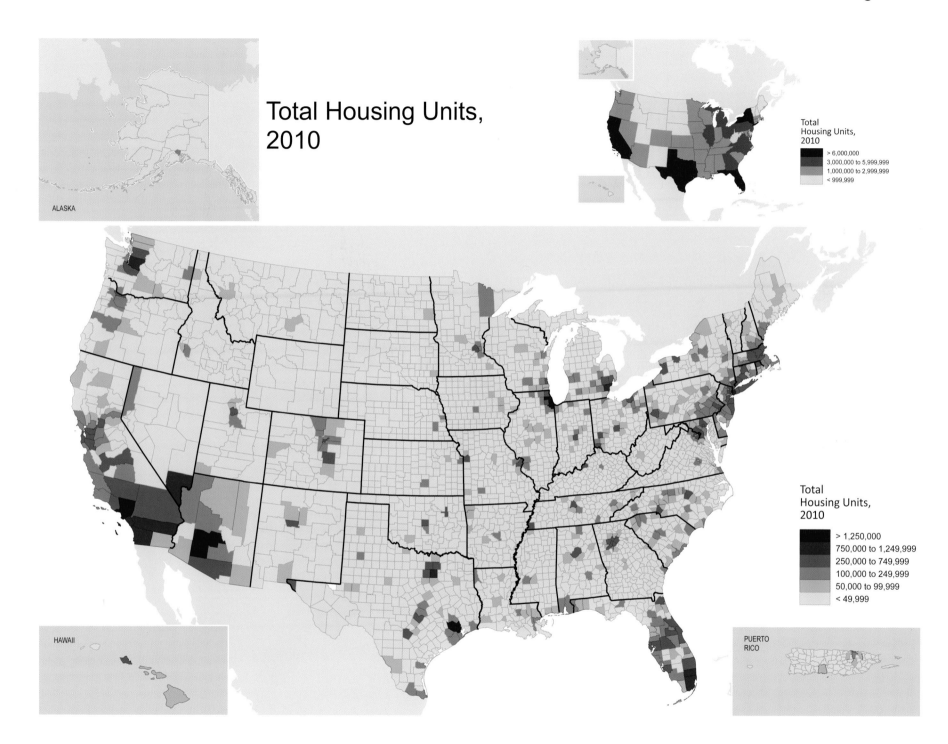

Total Housing Units,
2010

ALASKA

Total
Housing Units,
2010

> 6,000,000
3,000,000 to 5,999,999
1,000,000 to 2,999,999
< 999,999

Total
Housing Units,
2010

> 1,250,000
750,000 to 1,249,999
250,000 to 749,999
100,000 to 249,999
50,000 to 99,999
< 49,999

HAWAII

PUERTO
RICO

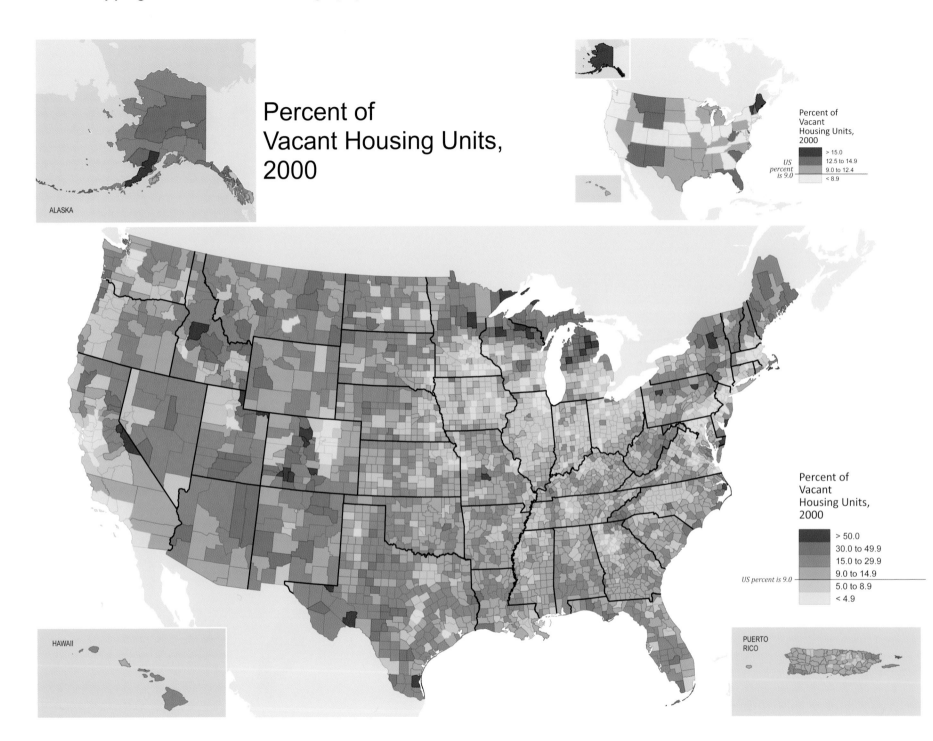

Percent of
Vacant Housing Units,
2000

ALASKA

Percent of
Vacant
Housing Units,
2000

> 15.0
12.5 to 14.9
US   9.0 to 12.4
percent
is 9.0   < 8.9

Percent of
Vacant
Housing Units,
2000

> 50.0
30.0 to 49.9
15.0 to 29.9
US percent is 9.0 — 9.0 to 14.9
5.0 to 8.9
< 4.9

HAWAII

PUERTO
RICO

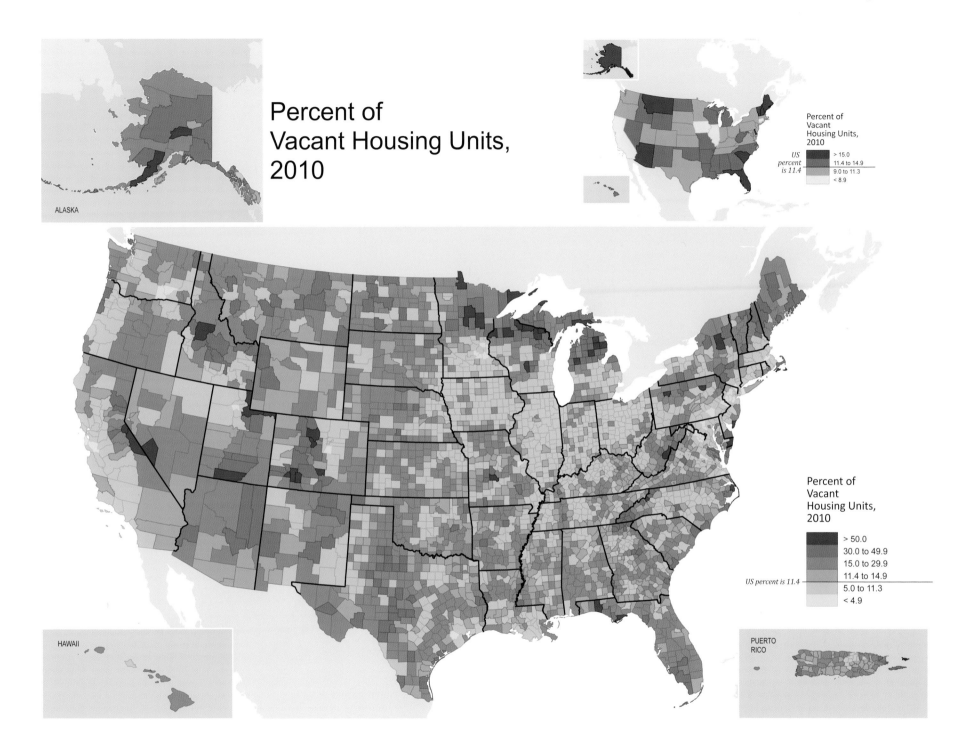

Percent of
Vacant Housing Units,
2010

ALASKA

Percent of
Vacant
Housing Units,
2010

US
percent
is 11.4

> 15.0
11.4 to 14.9
9.0 to 11.3
< 8.9

Percent of
Vacant
Housing Units,
2010

> 50.0
30.0 to 49.9
15.0 to 29.9
11.4 to 14.9
US percent is 11.4
5.0 to 11.3
< 4.9

HAWAII

PUERTO
RICO

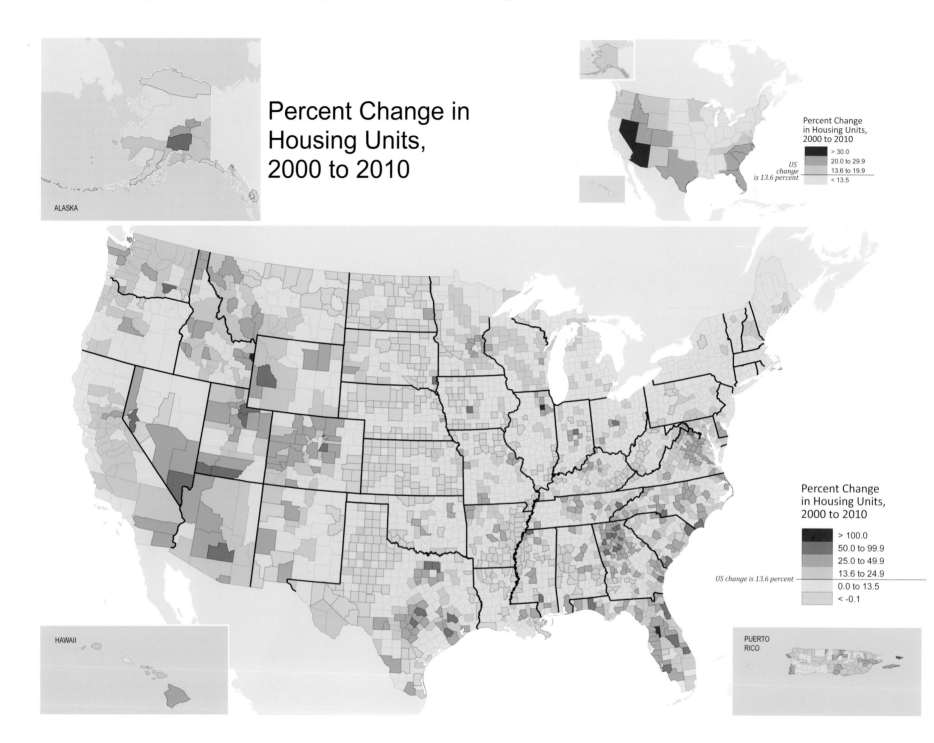

Percent Change in Housing Units, 2000 to 2010

ALASKA

Percent Change in Housing Units, 2000 to 2010

US change is 13.6 percent

> 30.0
20.0 to 29.9
13.6 to 19.9
< 13.5

Percent Change in Housing Units, 2000 to 2010

> 100.0
50.0 to 99.9
25.0 to 49.9
13.6 to 24.9
0.0 to 13.5
< -0.1

US change is 13.6 percent

HAWAII

PUERTO RICO

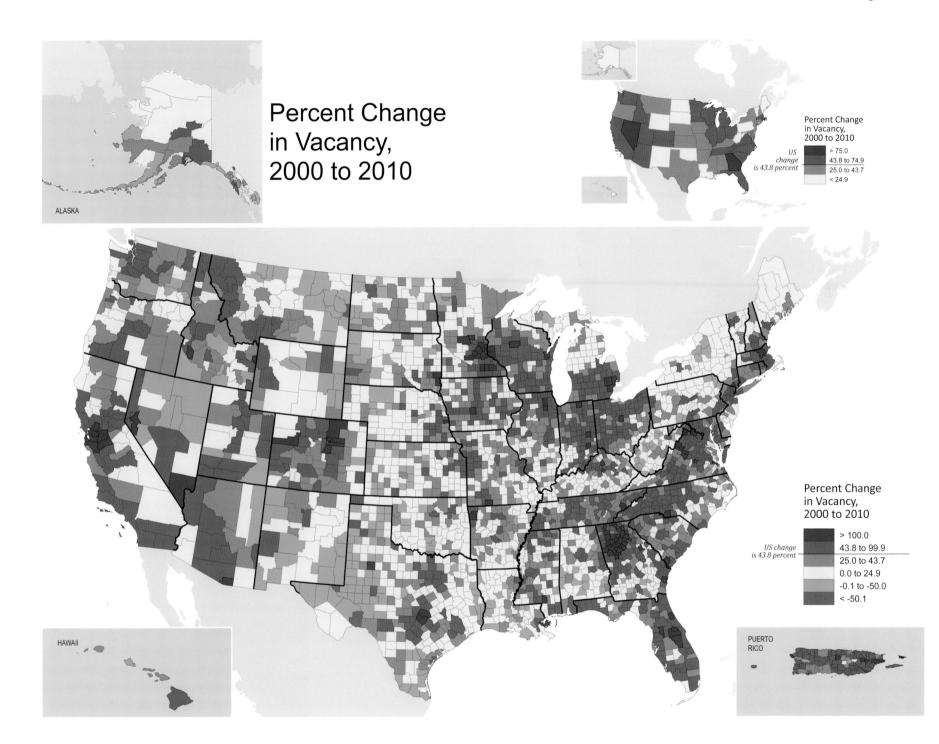

Percent Change
in Vacancy,
2000 to 2010

Percent Change
in Vacancy,
2000 to 2010

US change is 43.8 percent

> 75.0
43.8 to 74.9
25.0 to 43.7
< 24.9

ALASKA

Percent Change
in Vacancy,
2000 to 2010

US change
is 43.8 percent

> 100.0
43.8 to 99.9
25.0 to 43.7
0.0 to 24.9
-0.1 to -50.0
< -50.1

HAWAII

PUERTO
RICO

Section

3

Diversity

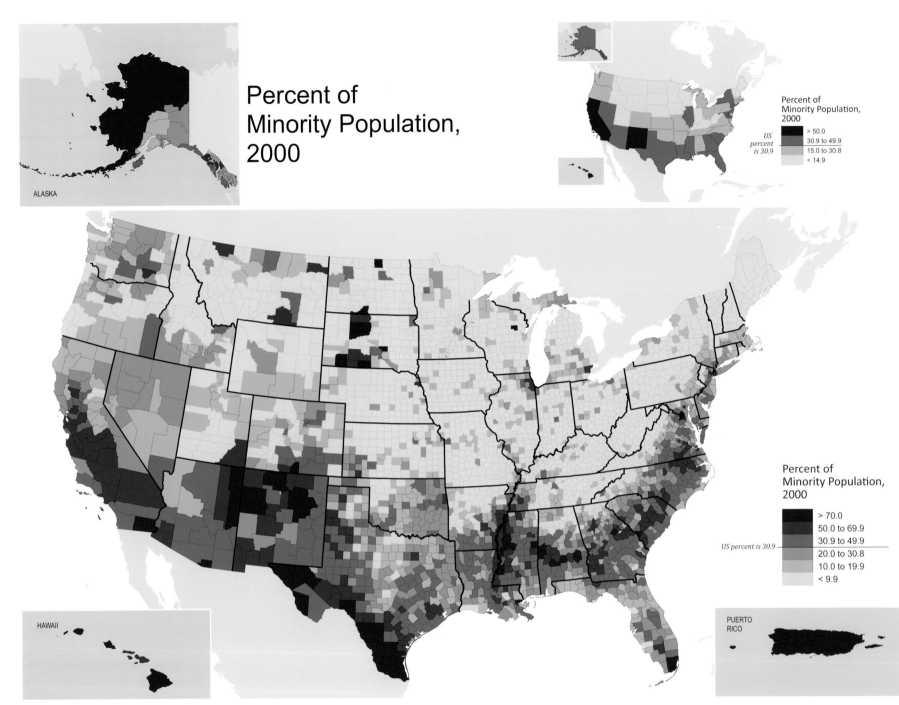

Percent of
Minority Population,
2000

ALASKA

Percent of
Minority Population,
2000

US
percent
is 30.9

> 50.0
30.9 to 49.9
15.0 to 30.8
< 14.9

Percent of
Minority Population,
2000

> 70.0
50.0 to 69.9
30.9 to 49.9
US percent is 30.9 —
20.0 to 30.8
10.0 to 19.9
< 9.9

HAWAII

PUERTO
RICO

Percent of Minority Population, 2010

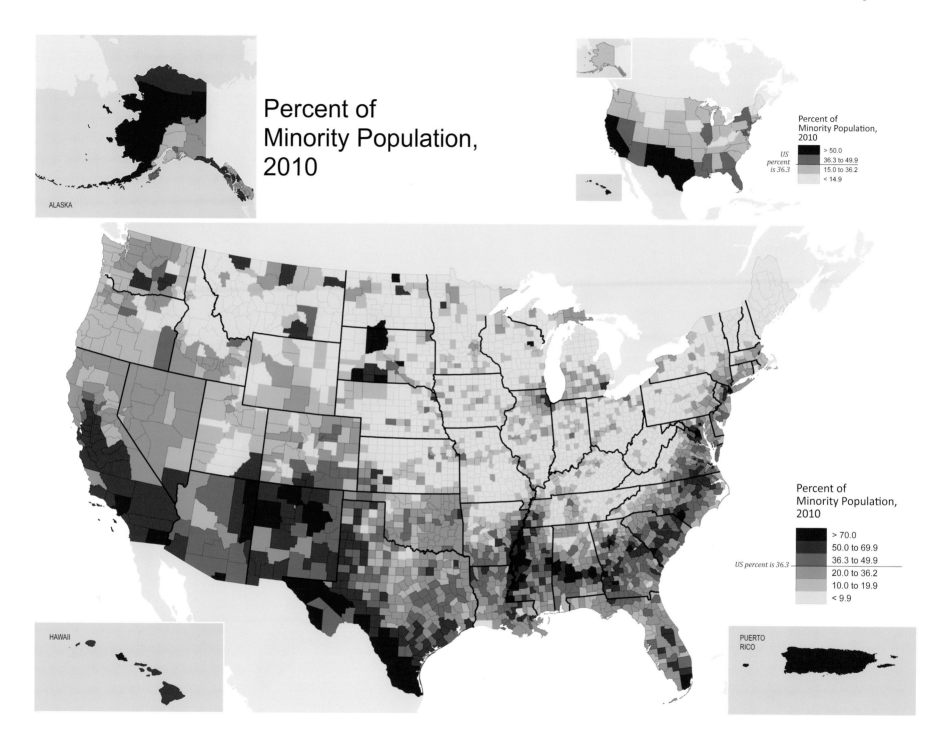

ALASKA

Percent of Minority Population, 2010

US percent is 36.3

> 50.0
36.3 to 49.9
15.0 to 36.2
< 14.9

Percent of Minority Population, 2010

> 70.0
50.0 to 69.9
36.3 to 49.9
20.0 to 36.2
10.0 to 19.9
< 9.9

US percent is 36.3

HAWAII

PUERTO RICO

 The minority population is composed of the total population less non-Hispanic whites.

Diversity Index, 2000

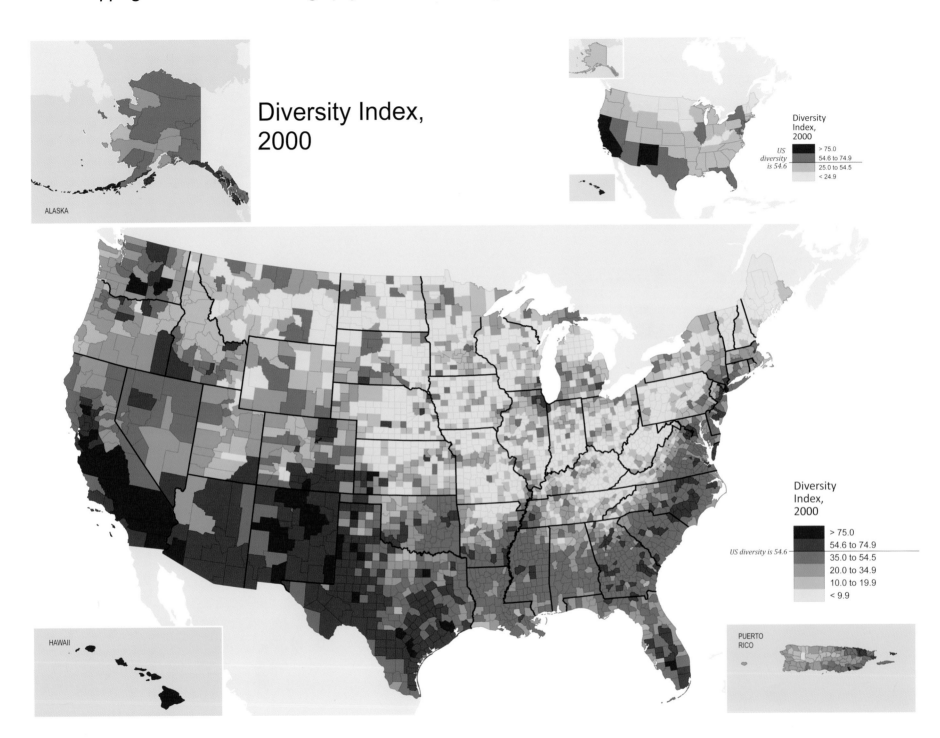

ALASKA

Diversity Index, 2000

- > 75.0
- 54.6 to 74.9
- 25.0 to 54.5
- < 24.9

US diversity is 54.6

Diversity Index, 2000

- > 75.0
- 54.6 to 74.9
- 35.0 to 54.5
- 20.0 to 34.9
- 10.0 to 19.9
- < 9.9

US diversity is 54.6

HAWAII

PUERTO RICO

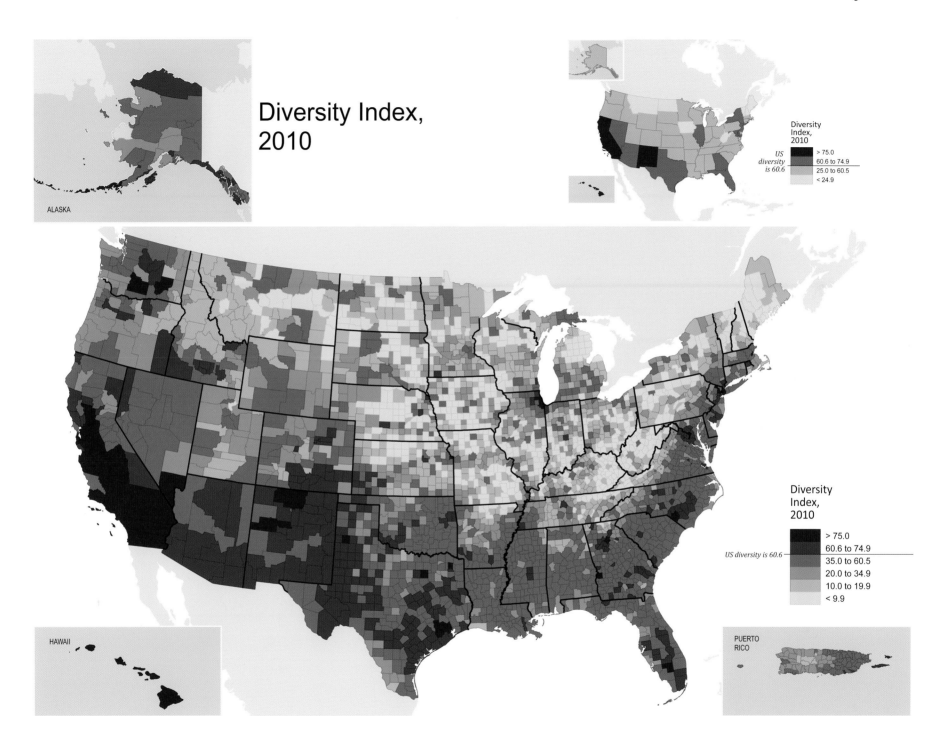

Diversity Index,
2010

ALASKA

Diversity
Index,
2010

US
diversity
is 60.6

> 75.0
60.6 to 74.9
25.0 to 60.5
< 24.9

Diversity
Index,
2010

US diversity is 60.6 —

> 75.0
60.6 to 74.9
35.0 to 60.5
20.0 to 34.9
10.0 to 19.9
< 9.9

HAWAII

PUERTO
RICO

Esri's Diversity Index represents the likelihood that two persons, chosen at random from the same area, belong to different race or ethnic groups. If an area's entire population belongs to one race group and one ethnic group, then an area has zero diversity. An area's diversity index increases to 100 when the population is evenly divided into two or more race/ethnic groups.

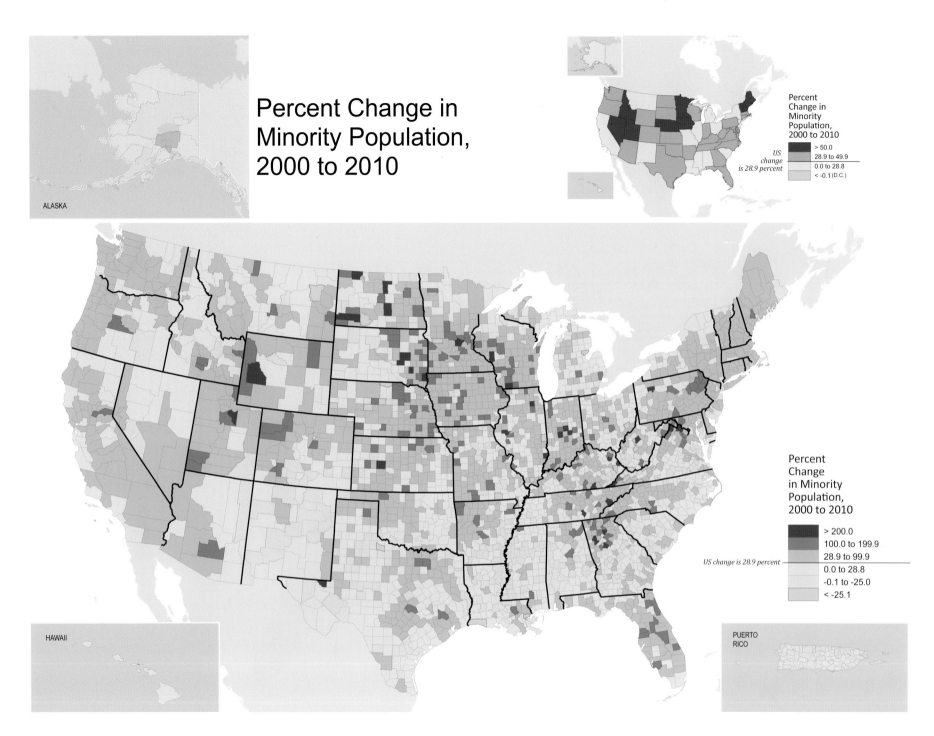

Percent Change in Minority Population, 2000 to 2010

ALASKA

Percent Change in Minority Population, 2000 to 2010

US change is 28.9 percent

> 50.0
28.9 to 49.9
0.0 to 28.8
< -0.1 (D.C.)

Percent Change in Minority Population, 2000 to 2010

> 200.0
100.0 to 199.9
28.9 to 99.9
US change is 28.9 percent
0.0 to 28.8
-0.1 to -25.0
< -25.1

HAWAII

PUERTO RICO

The minority population is composed of the total population less non-Hispanic whites.

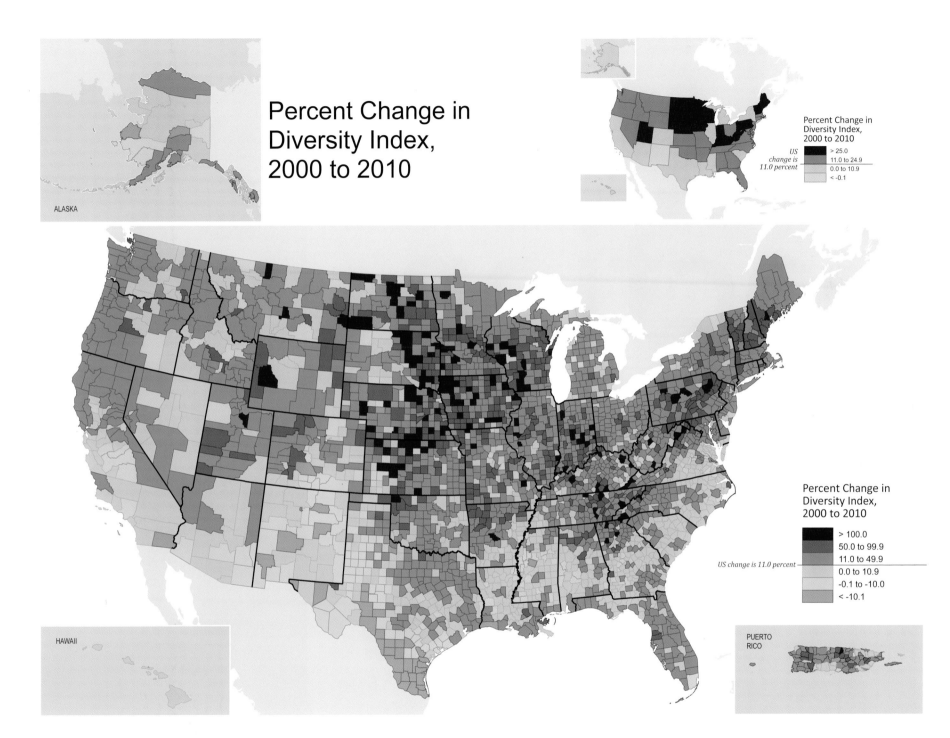

Percent Change in Diversity Index, 2000 to 2010

ALASKA

HAWAII

PUERTO RICO

Percent Change in Diversity Index, 2000 to 2010

US change is 11.0 percent

> 25.0
11.0 to 24.9
0.0 to 10.9
< -0.1

US change is 11.0 percent

Percent Change in Diversity Index, 2000 to 2010

> 100.0
50.0 to 99.9
11.0 to 49.9
0.0 to 10.9
-0.1 to -10.0
< -10.1

Esri's Diversity Index represents the likelihood that two persons, chosen at random from the same area, belong to different race or ethnic groups. If an area's entire population belongs to one race group and one ethnic group, then an area has zero diversity. An area's diversity index increases to 100 when the population is evenly divided into two or more race/ethnic groups.

Section

4

Non-Hispanic White

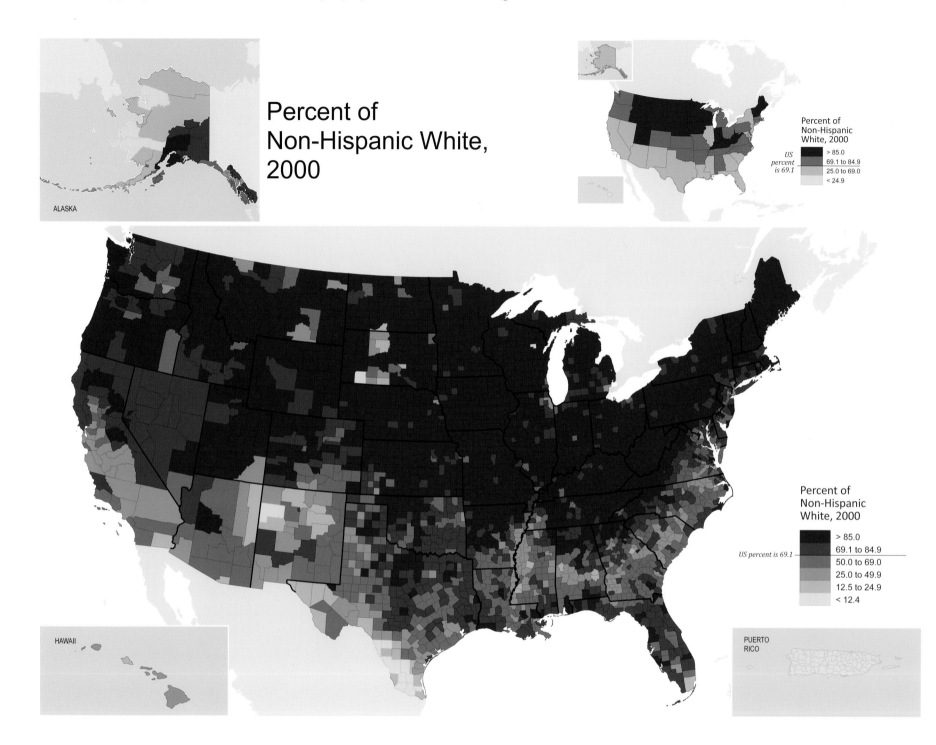

Percent of
Non-Hispanic White,
2000

ALASKA

HAWAII

PUERTO
RICO

Percent of
Non-Hispanic
White, 2000

US percent is 69.1

■	> 85.0
■	69.1 to 84.9
■	50.0 to 69.0
■	25.0 to 49.9
■	12.5 to 24.9
■	< 12.4

Percent of
Non-Hispanic
White, 2000

US percent is 69.1

■	> 85.0
■	69.1 to 84.9
■	25.0 to 69.0
■	< 24.9

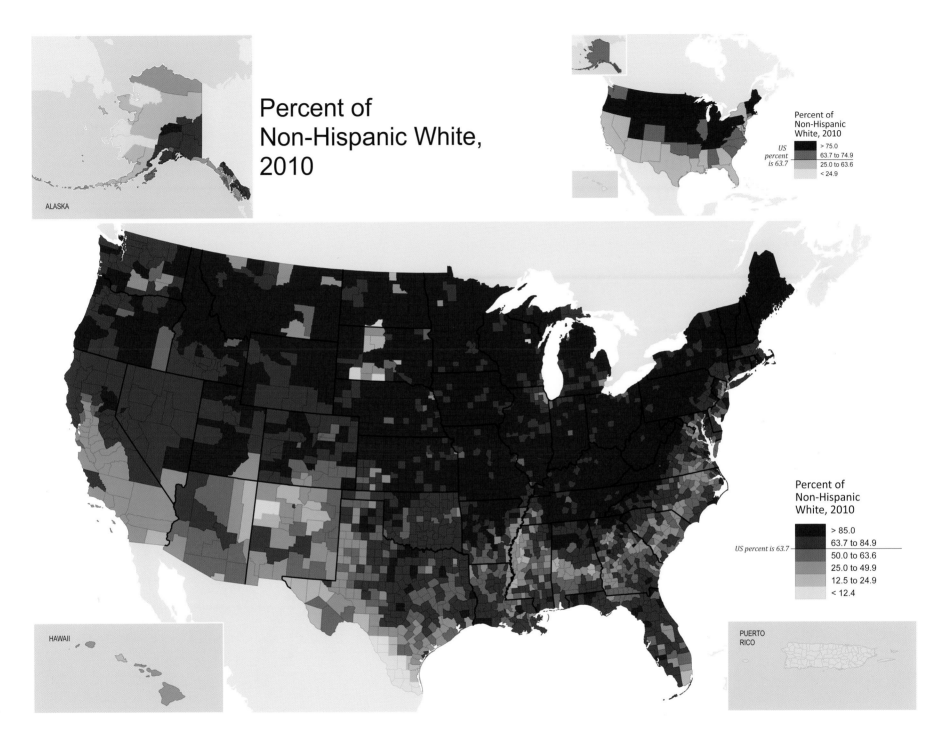

Percent of Non-Hispanic White, 2010

ALASKA

Percent of Non-Hispanic White, 2010

US percent is 63.7

> 75.0
63.7 to 74.9
25.0 to 63.6
< 24.9

Percent of Non-Hispanic White, 2010

US percent is 63.7

> 85.0
63.7 to 84.9
50.0 to 63.6
25.0 to 49.9
12.5 to 24.9
< 12.4

HAWAII

PUERTO RICO

 These maps show the share of the population that chose only one race category and chose non-Hispanic ethnicity.

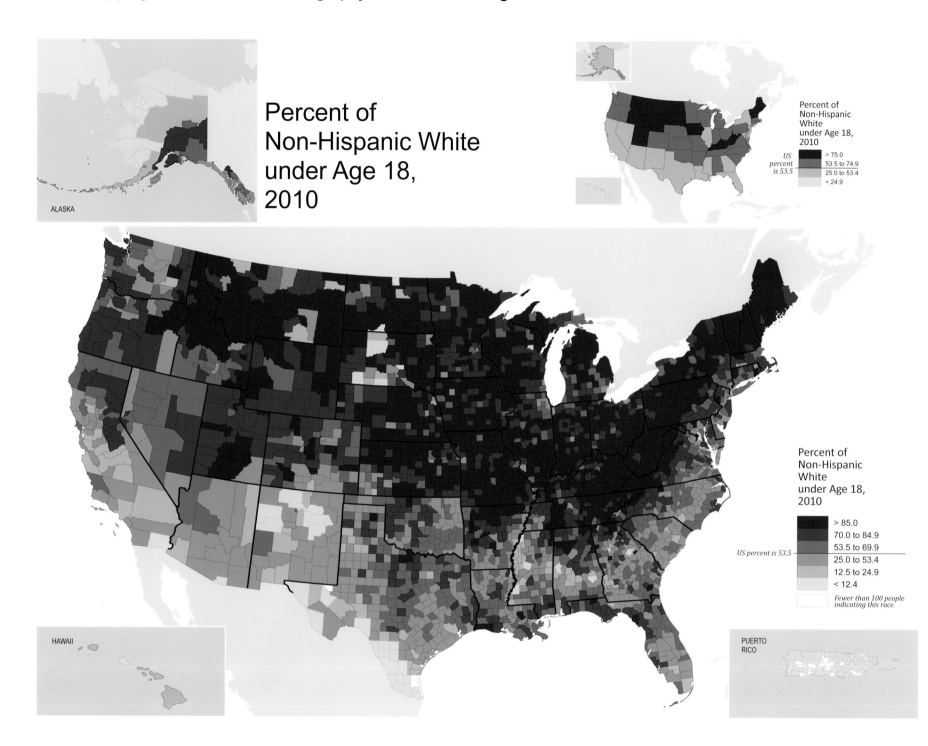

Percent of
Non-Hispanic White
under Age 18,
2010

ALASKA

Percent of
Non-Hispanic
White
under Age 18,
2010

US
percent
is 53.5

> 75.0
53.5 to 74.9
25.0 to 53.4
< 24.9

Percent of
Non-Hispanic
White
under Age 18,
2010

> 85.0
70.0 to 84.9
53.5 to 69.9

US percent is 53.5

25.0 to 53.4
12.5 to 24.9
< 12.4

*Fewer than 100 people
indicating this race.*

HAWAII

PUERTO
RICO

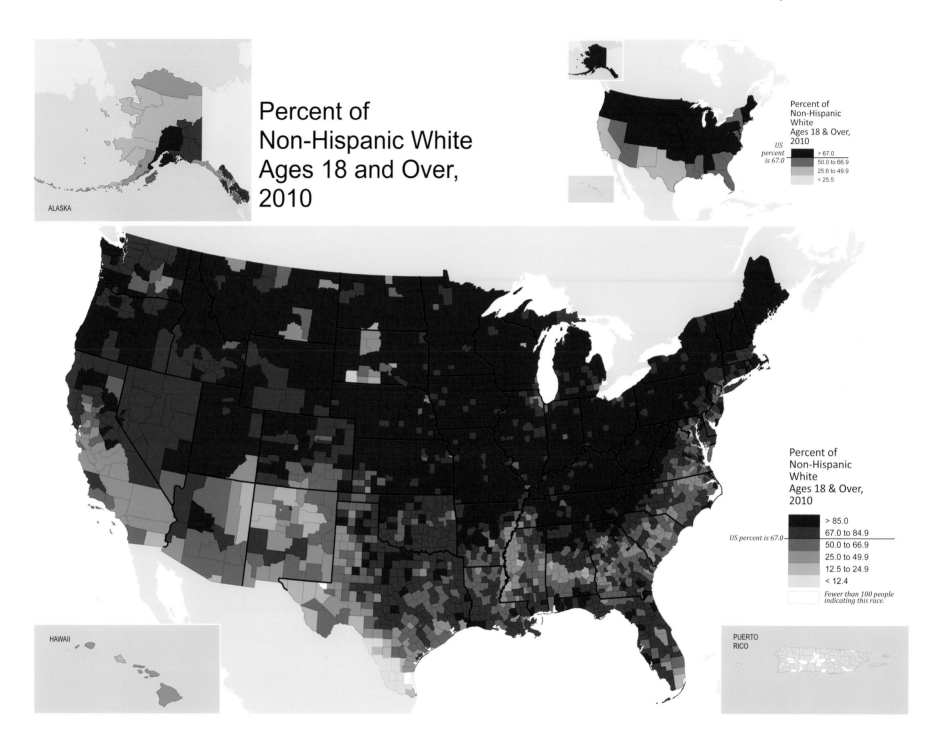

Percent of
Non-Hispanic White
Ages 18 and Over,
2010

ALASKA

Percent of
Non-Hispanic
White
Ages 18 & Over,
2010

US percent is 67.0

> 67.0
50.0 to 66.9
25.6 to 49.9
< 25.5

Percent of
Non-Hispanic
White
Ages 18 & Over,
2010

> 85.0
67.0 to 84.9
US percent is 67.0 —
50.0 to 66.9
25.0 to 49.9
12.5 to 24.9
< 12.4
Fewer than 100 people
indicating this race.

HAWAII

PUERTO
RICO

These maps show the share of the population that chose only one race category and chose non-Hispanic ethnicity.

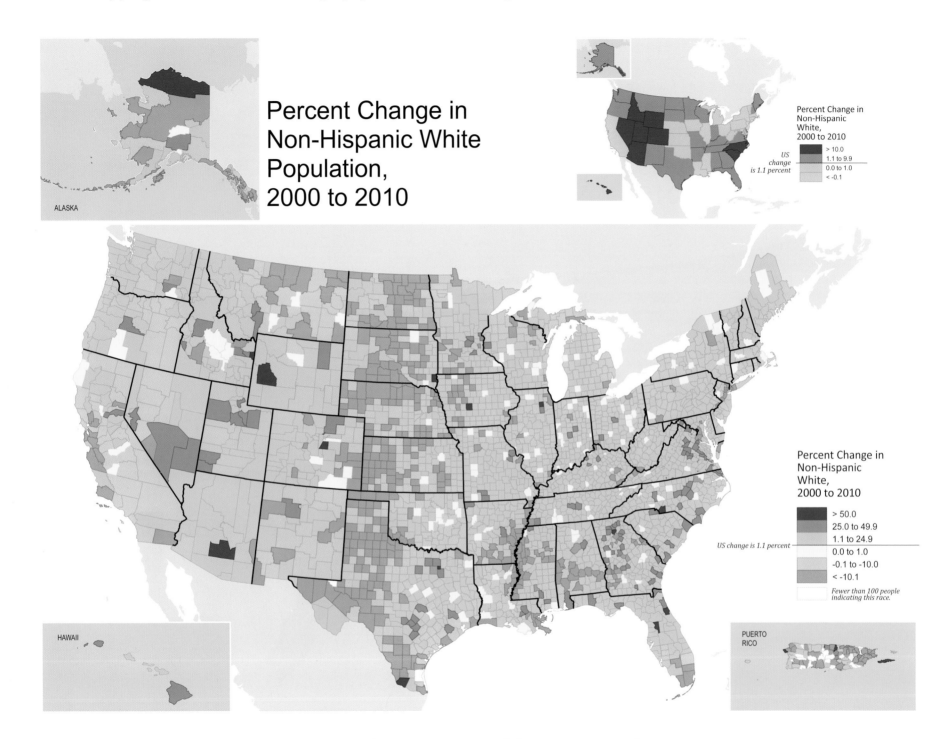

Percent Change in
Non-Hispanic White
Population,
2000 to 2010

ALASKA

Percent Change in
Non-Hispanic
White,
2000 to 2010

US
change
is 1.1 percent

> 10.0
1.1 to 9.9
0.0 to 1.0
< -0.1

Percent Change in
Non-Hispanic
White,
2000 to 2010

> 50.0
25.0 to 49.9
1.1 to 24.9
US change is 1.1 percent —— 0.0 to 1.0
-0.1 to -10.0
< -10.1

*Fewer than 100 people
indicating this race.*

HAWAII

PUERTO
RICO

These maps show the share of the population that chose only one race category and chose non-Hispanic ethnicity.

Black or
African American

Percent of Black or African American, 2000

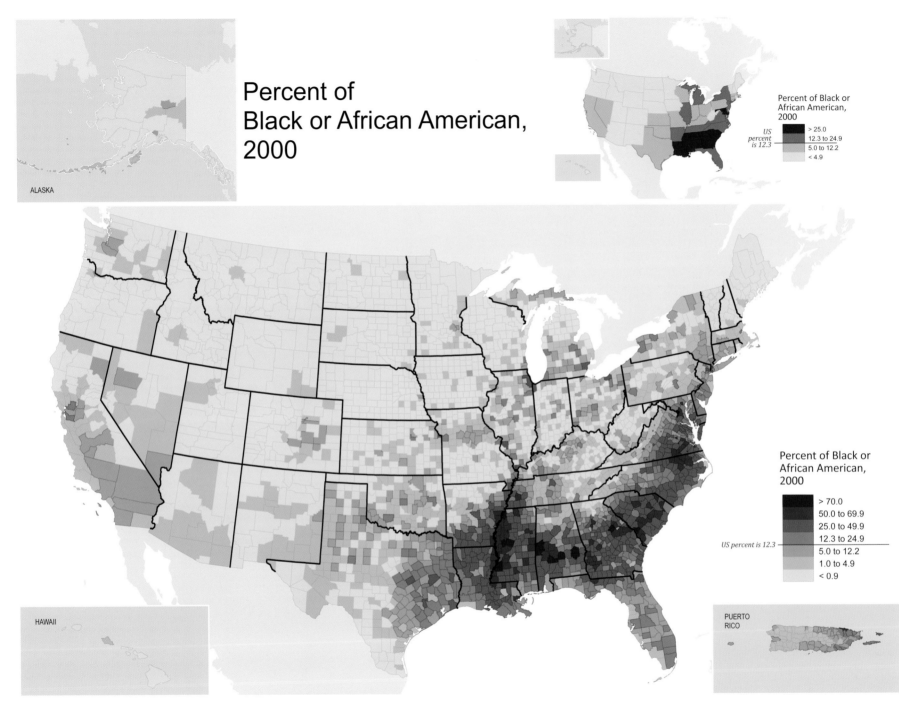

ALASKA

Percent of Black or African American, 2000

US percent is 12.3

| > 25.0 |
| 12.3 to 24.9 |
| 5.0 to 12.2 |
| < 4.9 |

Percent of Black or African American, 2000

| > 70.0 |
| 50.0 to 69.9 |
| 25.0 to 49.9 |
| 12.3 to 24.9 |
| 5.0 to 12.2 |
| 1.0 to 4.9 |
| < 0.9 |

US percent is 12.3

HAWAII

PUERTO RICO

Percent of
Black or African American,
2010

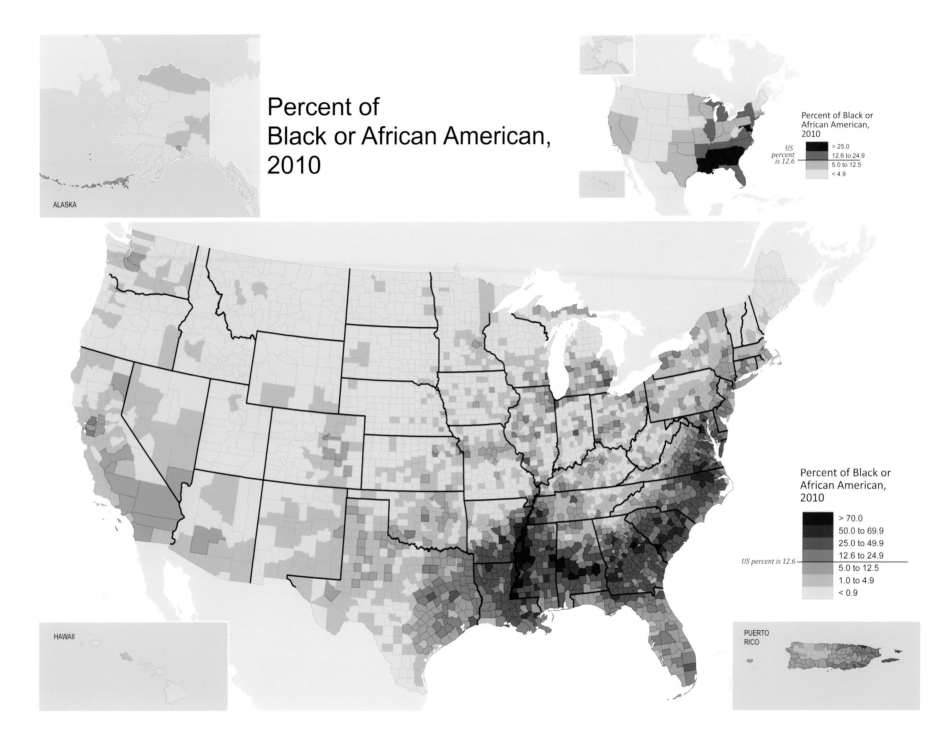

ALASKA

HAWAII

PUERTO RICO

Percent of Black or
African American,
2010

*US
percent
is 12.6*

> 25.0
12.6 to 24.9
5.0 to 12.5
< 4.9

Percent of Black or
African American,
2010

> 70.0
50.0 to 69.9
25.0 to 49.9
12.6 to 24.9

US percent is 12.6

5.0 to 12.5
1.0 to 4.9
< 0.9

 These maps show the share of the population that chose only one race category. This race category can be of either Hispanic or non-Hispanic ethnicity.

Percent of Black or African American under Age 18, 2010

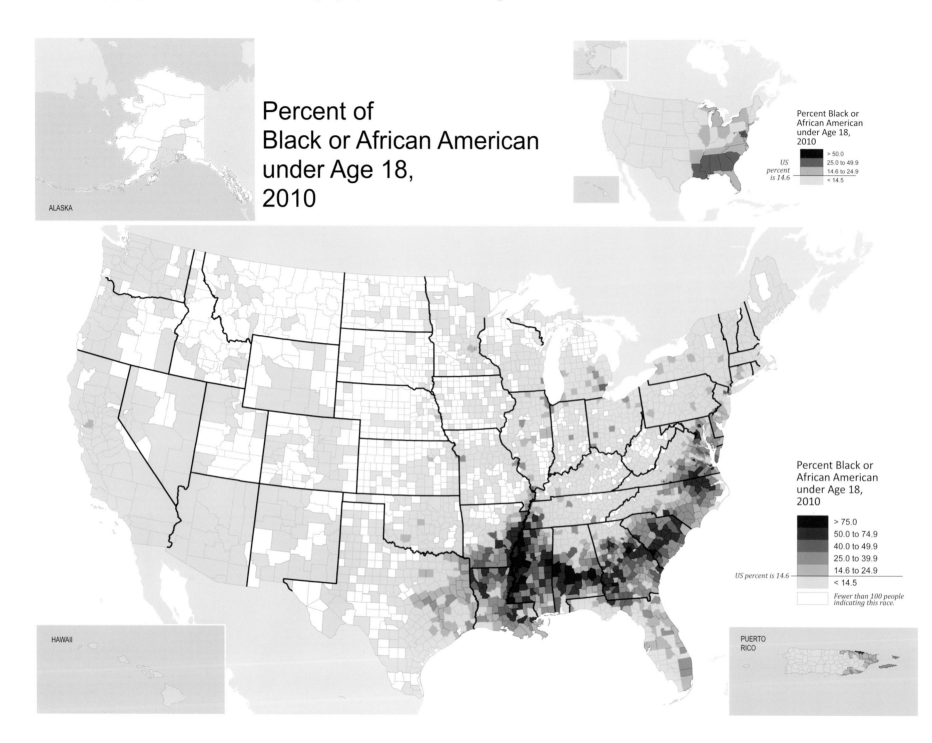

ALASKA

Percent Black or African American under Age 18, 2010

US percent is 14.6

- > 50.0
- 25.0 to 49.9
- 14.6 to 24.9
- < 14.5

Percent Black or African American under Age 18, 2010

- > 75.0
- 50.0 to 74.9
- 40.0 to 49.9
- 25.0 to 39.9
- 14.6 to 24.9
- US percent is 14.6
- < 14.5
- Fewer than 100 people indicating this race.

HAWAII

PUERTO RICO

Percent of Black or African American Ages 18 and Over, 2010

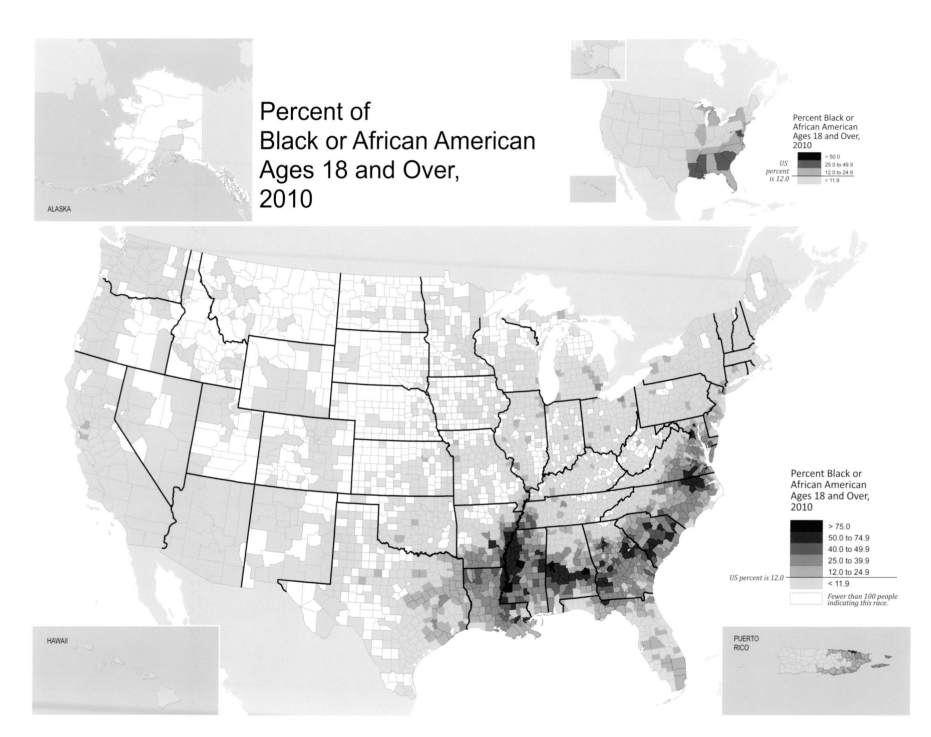

ALASKA

Percent Black or African American Ages 18 and Over, 2010

US percent is 12.0

> 50.0
25.0 to 49.9
12.0 to 24.9
< 11.9

Percent Black or African American Ages 18 and Over, 2010

> 75.0
50.0 to 74.9
40.0 to 49.9
25.0 to 39.9
12.0 to 24.9
US percent is 12.0
< 11.9

Fewer than 100 people indicating this race.

HAWAII

PUERTO RICO

 These maps show the share of the population that chose only one race category. This race category can be of either Hispanic or non-Hispanic ethnicity.

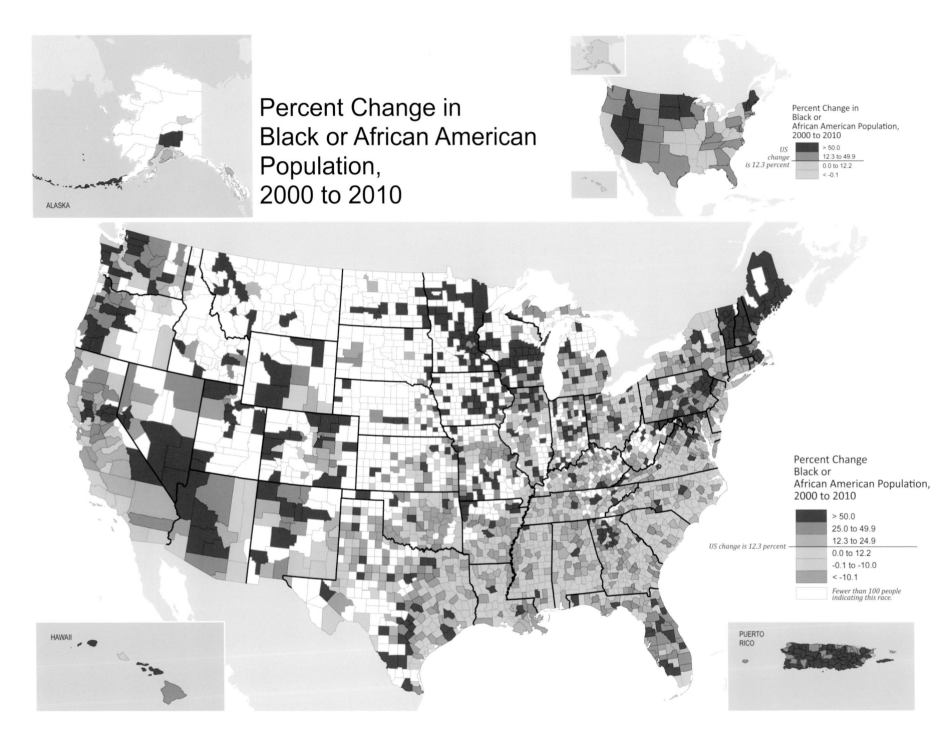

ALASKA

Percent Change in
Black or African American
Population,
2000 to 2010

Percent Change in
Black or
African American Population,
2000 to 2010

US
change
is 12.3 percent

> 50.0
12.3 to 49.9
0.0 to 12.2
< -0.1

Percent Change
Black or
African American Population,
2000 to 2010

> 50.0
25.0 to 49.9
12.3 to 24.9
0.0 to 12.2
-0.1 to -10.0
< -10.1

US change is 12.3 percent

Fewer than 100 people
indicating this race.

HAWAII

PUERTO
RICO

These maps show the share of the population that chose only one race category. This race category can be of either Hispanic
or non-Hispanic ethnicity.

American Indian or Alaska Native

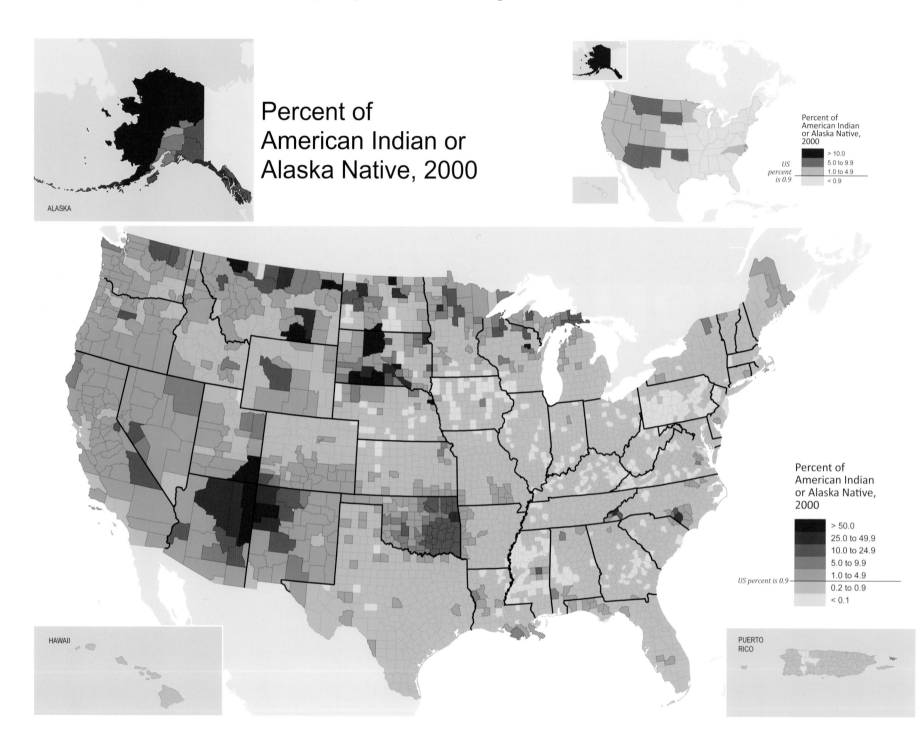

Percent of
American Indian or
Alaska Native, 2000

ALASKA

Percent of
American Indian
or Alaska Native,
2000

US
percent
is 0.9

> 10.0
5.0 to 9.9
1.0 to 4.9
< 0.9

Percent of
American Indian
or Alaska Native,
2000

> 50.0
25.0 to 49.9
10.0 to 24.9
5.0 to 9.9
1.0 to 4.9
US percent is 0.9 — 0.2 to 0.9
< 0.1

HAWAII

PUERTO
RICO

Percent of American Indian or Alaska Native, 2010

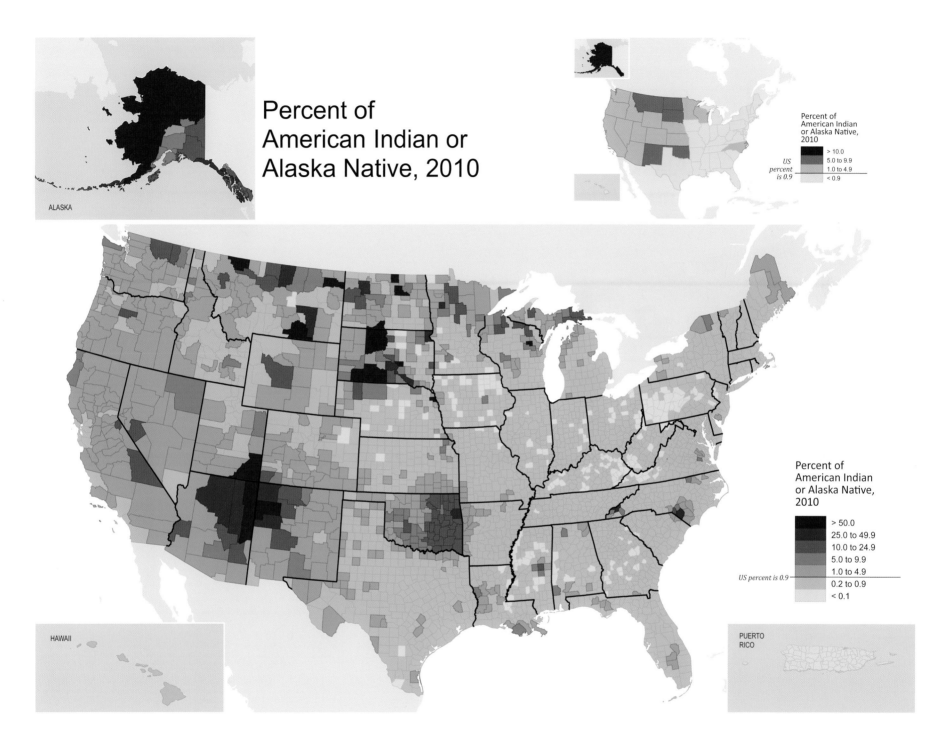

ALASKA

Percent of American Indian or Alaska Native, 2010

> 10.0
5.0 to 9.9
1.0 to 4.9
< 0.9

US percent is 0.9

Percent of American Indian or Alaska Native, 2010

> 50.0
25.0 to 49.9
10.0 to 24.9
5.0 to 9.9
1.0 to 4.9
US percent is 0.9
0.2 to 0.9
< 0.1

HAWAII

PUERTO RICO

 These maps show the share of the population that chose only one race category. This race category can be of either Hispanic or non-Hispanic ethnicity.

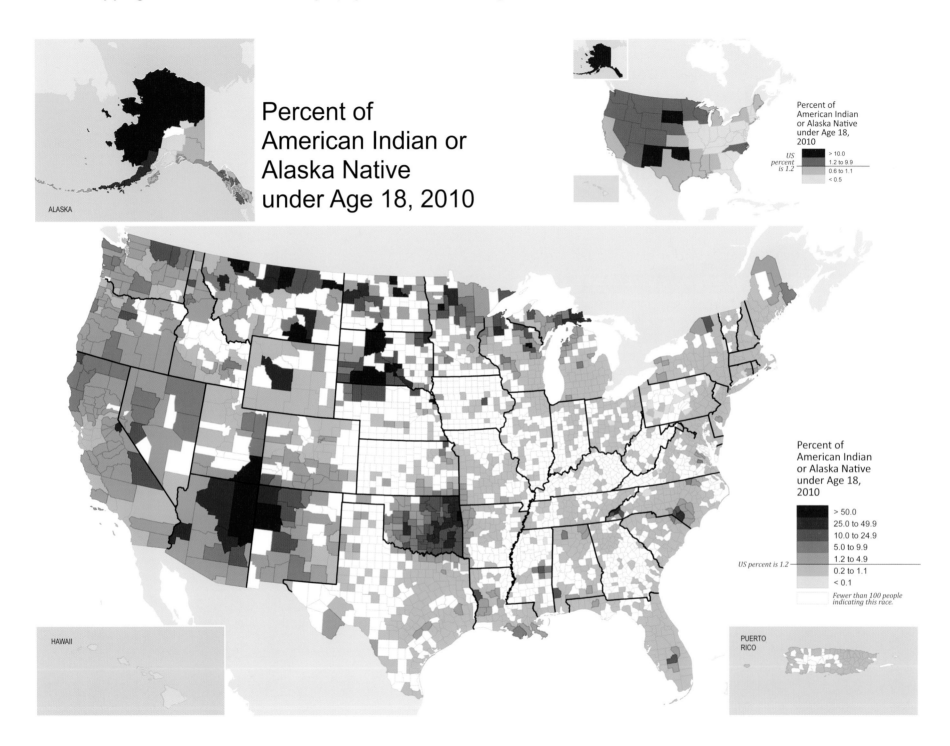

Percent of
American Indian or
Alaska Native
under Age 18, 2010

ALASKA

Percent of
American Indian
or Alaska Native
under Age 18,
2010

US percent is 1.2

> 10.0
1.2 to 9.9
0.6 to 1.1
< 0.5

Percent of
American Indian
or Alaska Native
under Age 18,
2010

> 50.0
25.0 to 49.9
10.0 to 24.9
5.0 to 9.9
1.2 to 4.9
0.2 to 1.1
< 0.1

US percent is 1.2

Fewer than 100 people
indicating this race.

HAWAII

PUERTO
RICO

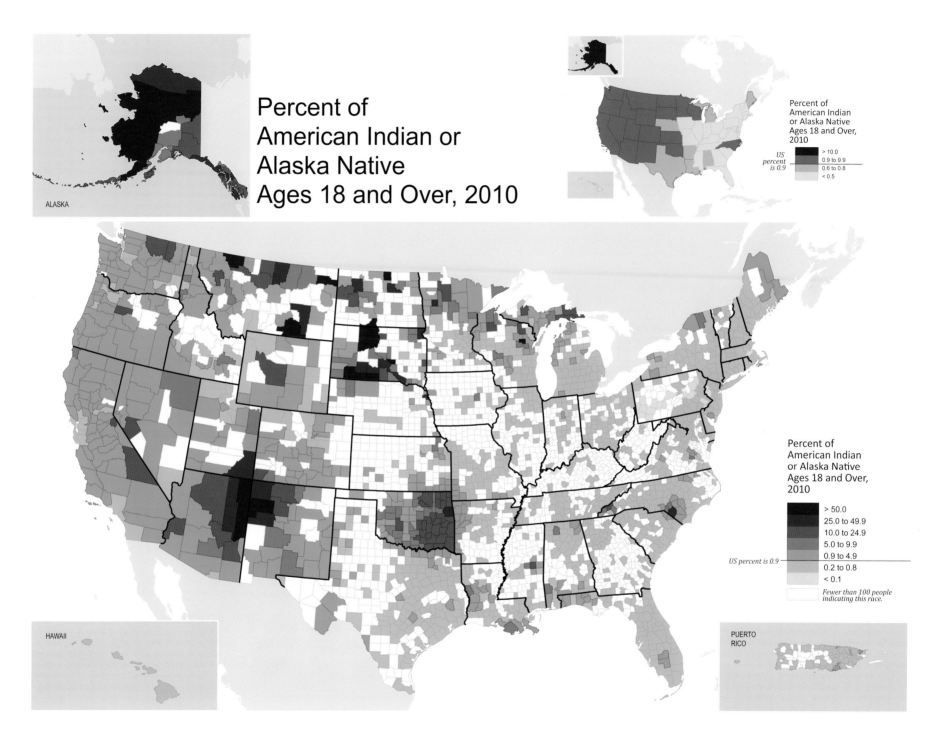

Percent of
American Indian or
Alaska Native
Ages 18 and Over, 2010

ALASKA

Percent of
American Indian
or Alaska Native
Ages 18 and Over,
2010

US percent is 0.9

> 10.0
0.9 to 9.9
0.6 to 0.8
< 0.5

Percent of
American Indian
or Alaska Native
Ages 18 and Over,
2010

> 50.0
25.0 to 49.9
10.0 to 24.9
5.0 to 9.9
0.9 to 4.9
0.2 to 0.8
< 0.1

US percent is 0.9

Fewer than 100 people
indicating this race.

HAWAII

PUERTO
RICO

 These maps show the share of the population that chose only one race category. This race category can be of either Hispanic or non-Hispanic ethnicity.

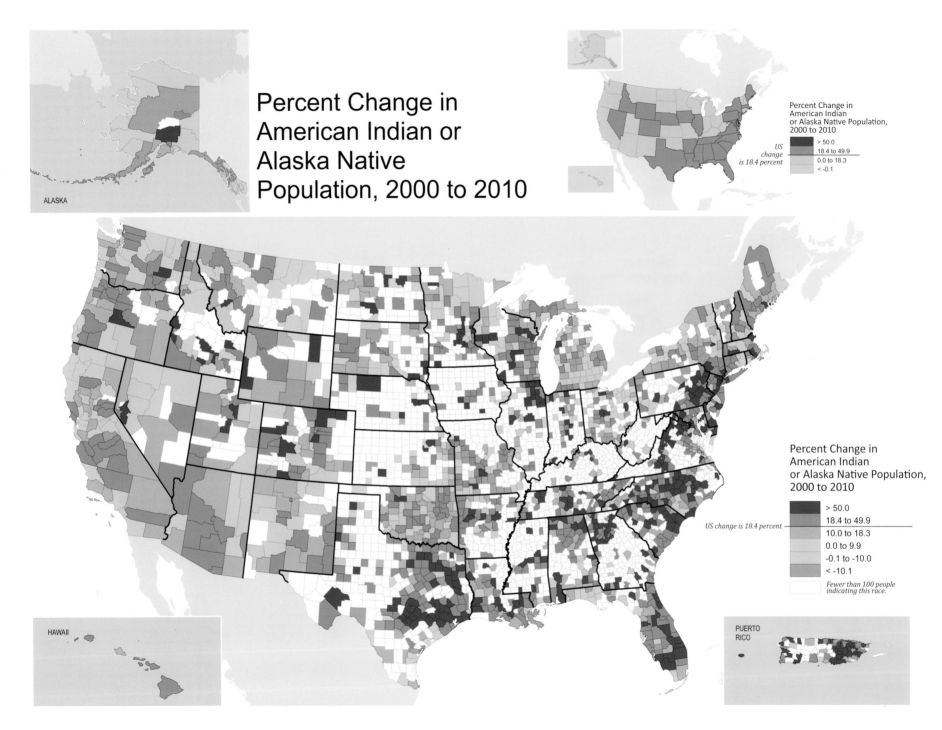

Percent Change in American Indian or Alaska Native Population, 2000 to 2010

ALASKA

Percent Change in American Indian or Alaska Native Population, 2000 to 2010

US change is 18.4 percent

> 50.0
18.4 to 49.9
0.0 to 18.3
< -0.1

Percent Change in American Indian or Alaska Native Population, 2000 to 2010

> 50.0
18.4 to 49.9
10.0 to 18.3
0.0 to 9.9
-0.1 to -10.0
< -10.1
Fewer than 100 people indicating this race.

US change is 18.4 percent

HAWAII

PUERTO RICO

These maps show the share of the population that chose only one race category. This race category can be of either Hispanic or non-Hispanic ethnicity.

Asian

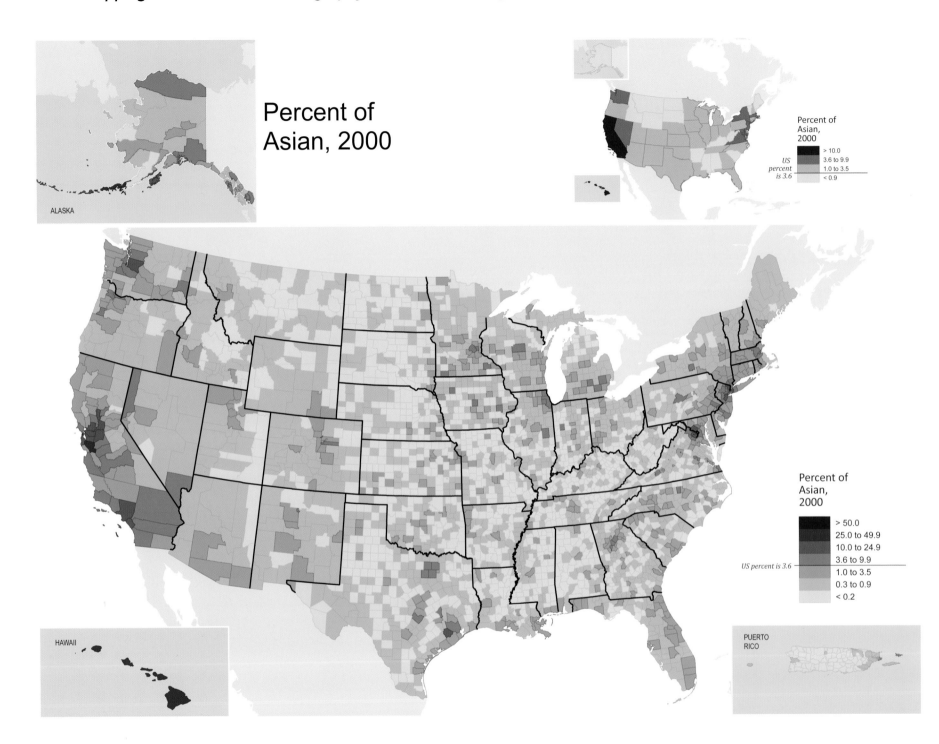

Percent of Asian, 2000

ALASKA

Percent of
Asian,
2000

*US
percent
is 3.6*

> 10.0
3.6 to 9.9
1.0 to 3.5
< 0.9

Percent of
Asian,
2000

> 50.0
25.0 to 49.9
10.0 to 24.9
3.6 to 9.9
US percent is 3.6
1.0 to 3.5
0.3 to 0.9
< 0.2

HAWAII

PUERTO
RICO

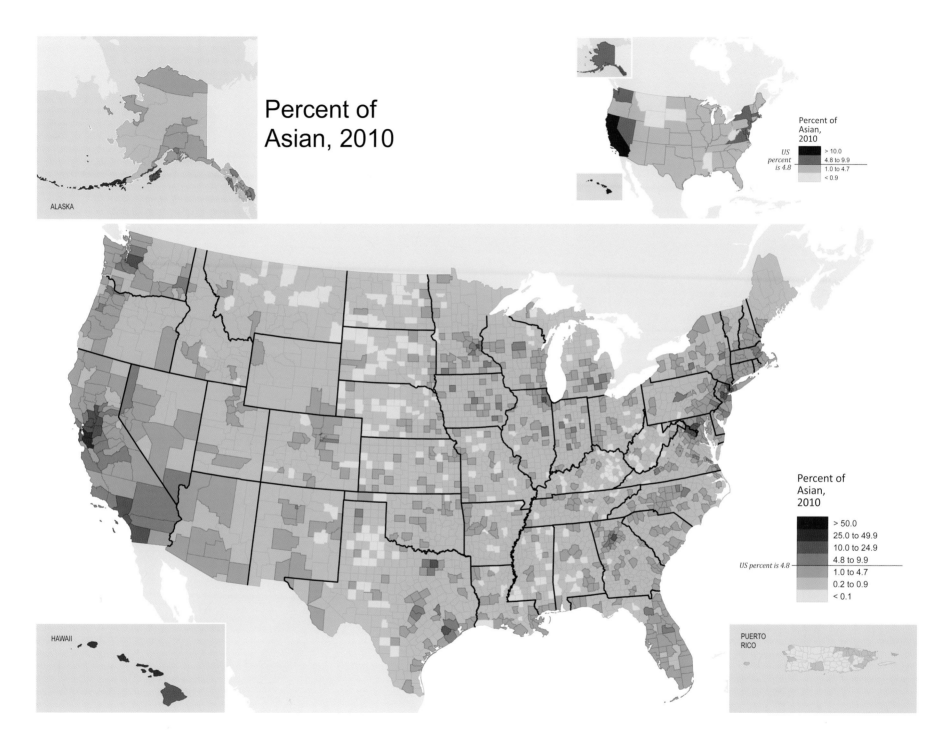

Percent of Asian, 2010

Percent of Asian, 2010

US percent is 4.8

> 10.0
4.8 to 9.9
1.0 to 4.7
< 0.9

ALASKA

HAWAII

PUERTO RICO

Percent of Asian, 2010

> 50.0
25.0 to 49.9
10.0 to 24.9
4.8 to 9.9
1.0 to 4.7
0.2 to 0.9
< 0.1

US percent is 4.8

These maps show the share of the population that chose only one race category. This race category can be of either Hispanic or non-Hispanic ethnicity.

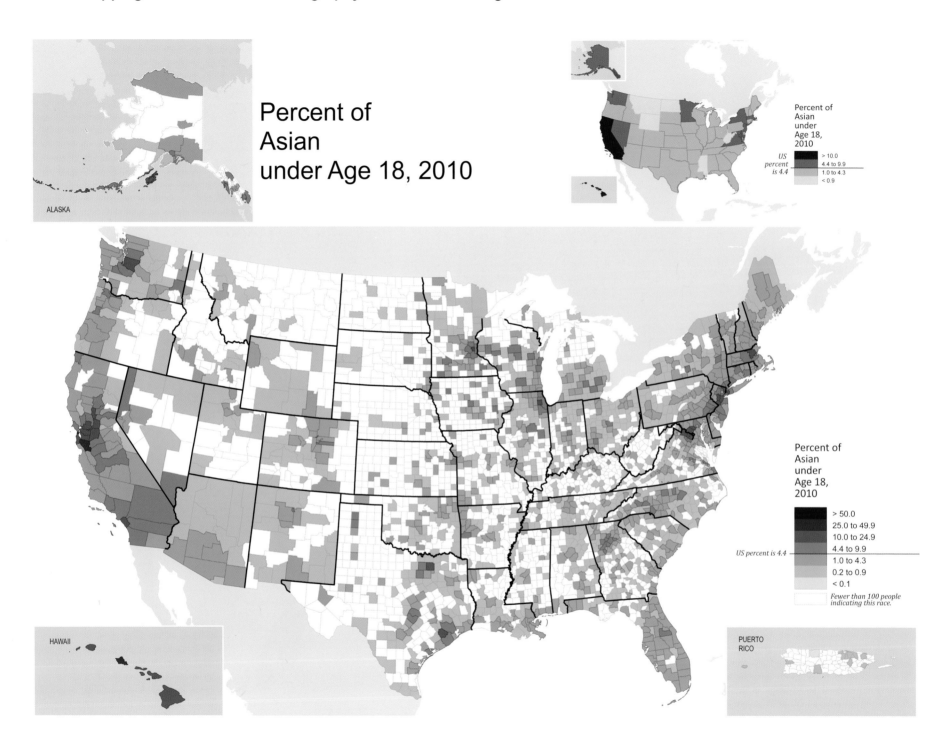

Percent of
Asian
under Age 18, 2010

ALASKA

Percent of
Asian
under
Age 18,
2010

US
percent
is 4.4

> 10.0
4.4 to 9.9
1.0 to 4.3
< 0.9

Percent of
Asian
under
Age 18,
2010

US percent is 4.4

> 50.0
25.0 to 49.9
10.0 to 24.9
4.4 to 9.9
1.0 to 4.3
0.2 to 0.9
< 0.1

*Fewer than 100 people
indicating this race.*

HAWAII

PUERTO
RICO

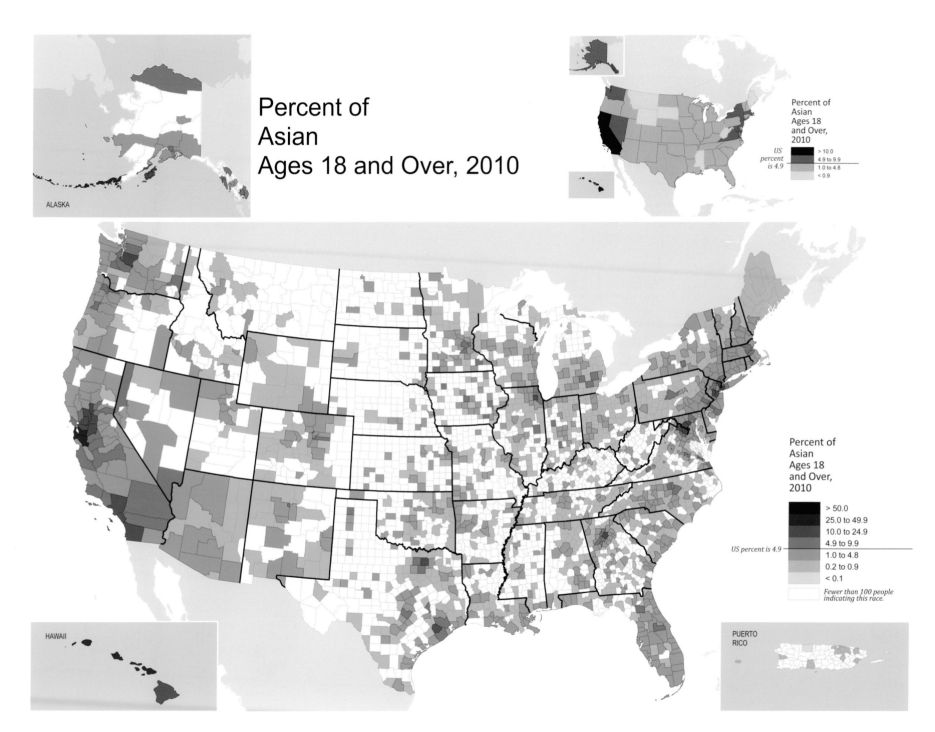

Percent of
Asian
Ages 18 and Over, 2010

ALASKA

Percent of
Asian
Ages 18
and Over,
2010

US
percent
is 4.9

> 10.0
4.9 to 9.9
1.0 to 4.8
< 0.9

Percent of
Asian
Ages 18
and Over,
2010

> 50.0
25.0 to 49.9
10.0 to 24.9
4.9 to 9.9
1.0 to 4.8
0.2 to 0.9
< 0.1

US percent is 4.9

Fewer than 100 people
indicating this race.

HAWAII

PUERTO
RICO

 These maps show the share of the population that chose only one race category. This race category can be of either Hispanic or non-Hispanic ethnicity.

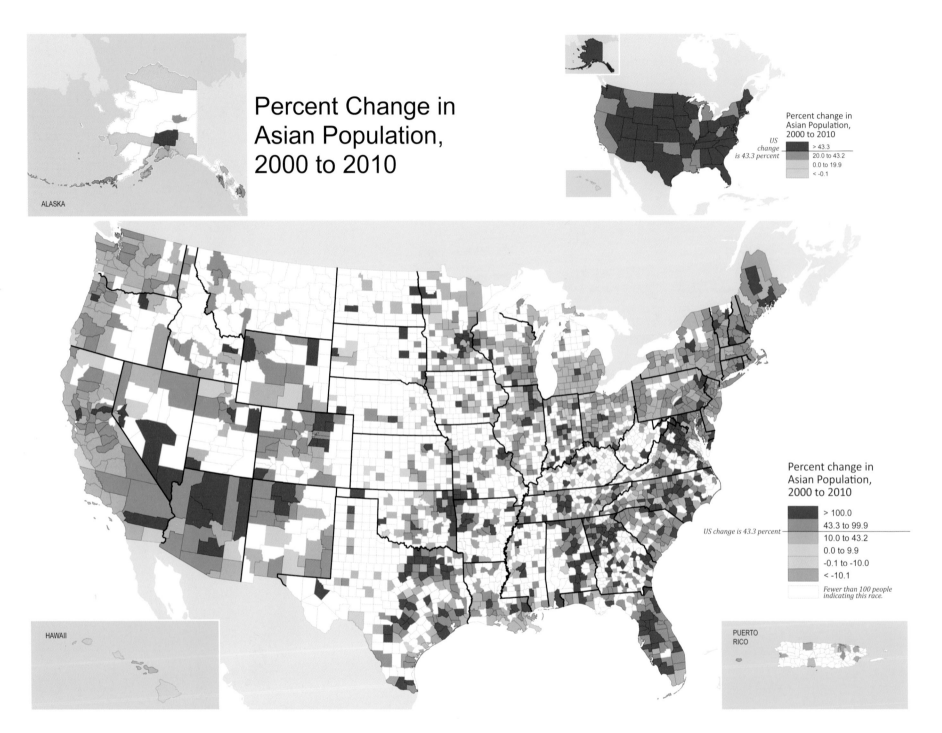

Percent Change in Asian Population, 2000 to 2010

ALASKA

Percent change in Asian Population, 2000 to 2010

US change is 43.3 percent

	> 43.3
	20.0 to 43.2
	0.0 to 19.9
	< -0.1

Percent change in Asian Population, 2000 to 2010

US change is 43.3 percent

	> 100.0
	43.3 to 99.9
	10.0 to 43.2
	0.0 to 9.9
	-0.1 to -10.0
	< -10.1
	Fewer than 100 people indicating this race.

HAWAII

PUERTO RICO

These maps show the share of the population that chose only one race category. This race category can be of either Hispanic or non-Hispanic ethnicity.

Native Hawaiian or Other Pacific Islander

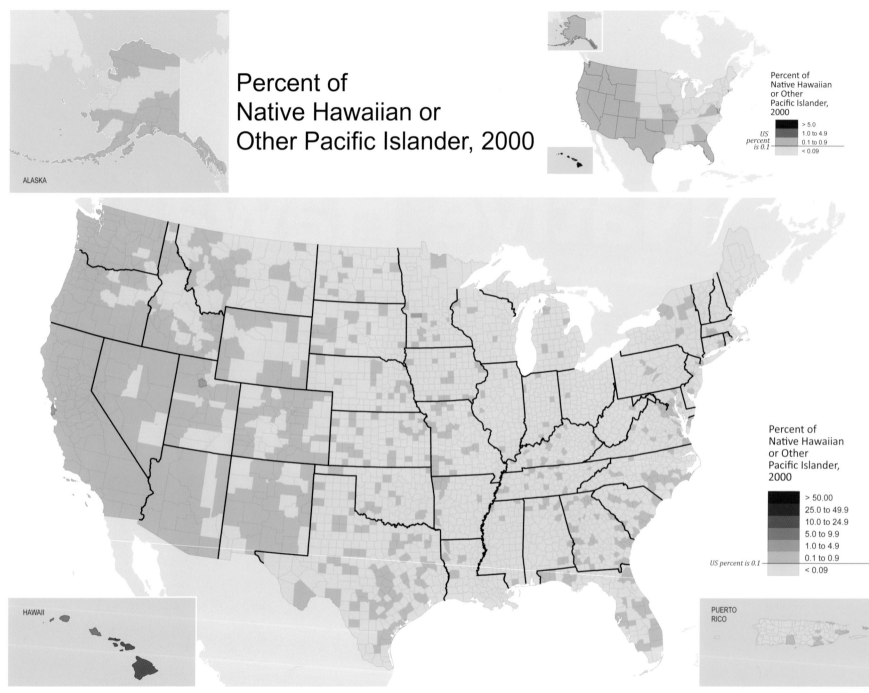

Percent of Native Hawaiian or Other Pacific Islander, 2000

ALASKA

Percent of
Native Hawaiian
or Other
Pacific Islander,
2000

US percent is 0.1

> 5.0
1.0 to 4.9
0.1 to 0.9
< 0.09

Percent of
Native Hawaiian
or Other
Pacific Islander,
2000

> 50.00
25.0 to 49.9
10.0 to 24.9
5.0 to 9.9
1.0 to 4.9
US percent is 0.1 — 0.1 to 0.9
< 0.09

HAWAII

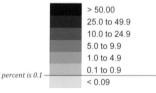

PUERTO
RICO

Percent of Native Hawaiian or Other Pacific Islander, 2010

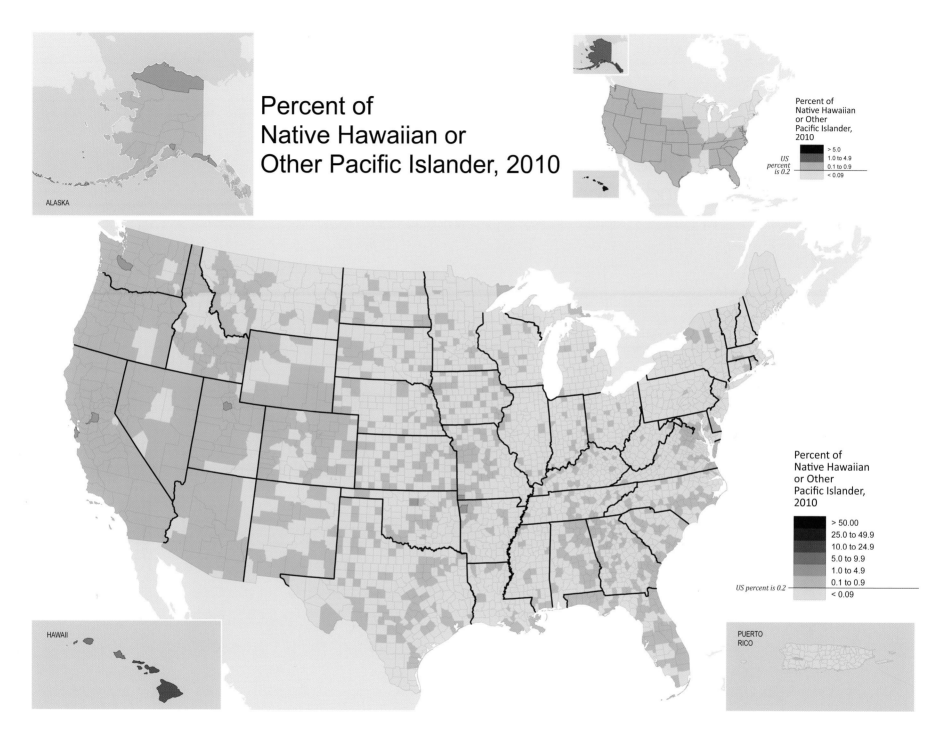

ALASKA

Percent of
Native Hawaiian
or Other
Pacific Islander,
2010

US
percent
is 0.2
| > 5.0 |
| 1.0 to 4.9 |
| 0.1 to 0.9 |
| < 0.09 |

Percent of
Native Hawaiian
or Other
Pacific Islander,
2010

| > 50.00 |
| 25.0 to 49.9 |
| 10.0 to 24.9 |
| 5.0 to 9.9 |
| 1.0 to 4.9 |
| 0.1 to 0.9 |
| < 0.09 |

US percent is 0.2

HAWAII

PUERTO RICO

 These maps show the share of the population that chose only one race category. This race category can be of either Hispanic or non-Hispanic ethnicity.

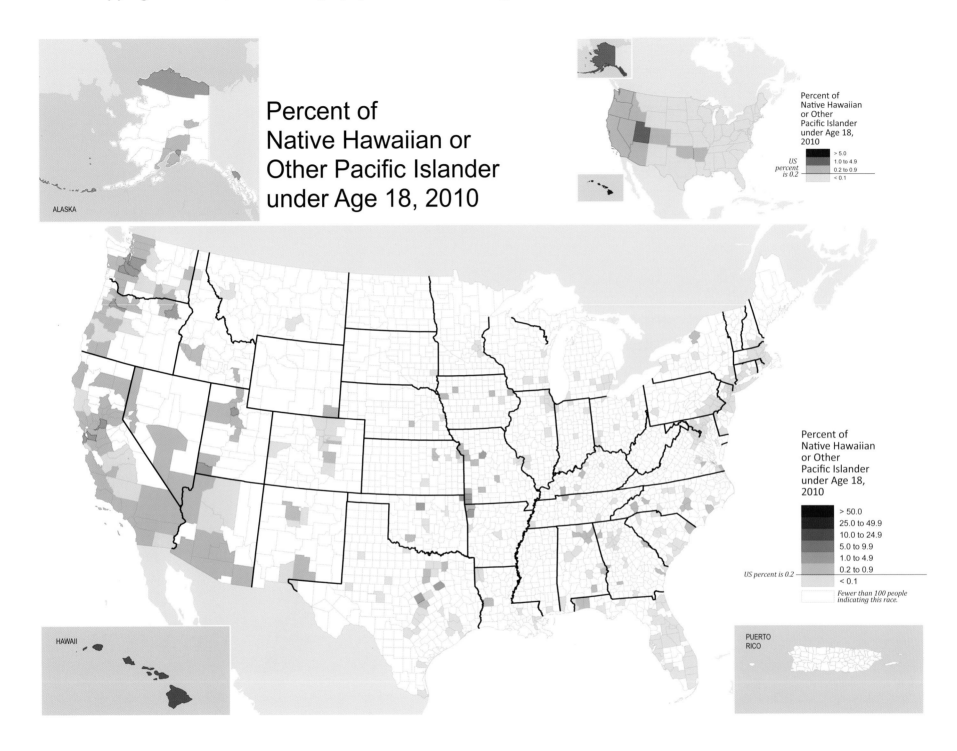

Percent of
Native Hawaiian or
Other Pacific Islander
under Age 18, 2010

ALASKA

Percent of
Native Hawaiian
or Other
Pacific Islander
under Age 18,
2010

US
percent
is 0.2

> 5.0
1.0 to 4.9
0.2 to 0.9
< 0.1

Percent of
Native Hawaiian
or Other
Pacific Islander
under Age 18,
2010

> 50.0
25.0 to 49.9
10.0 to 24.9
5.0 to 9.9
1.0 to 4.9
0.2 to 0.9

US percent is 0.2

< 0.1

*Fewer than 100 people
indicating this race.*

HAWAII

PUERTO
RICO

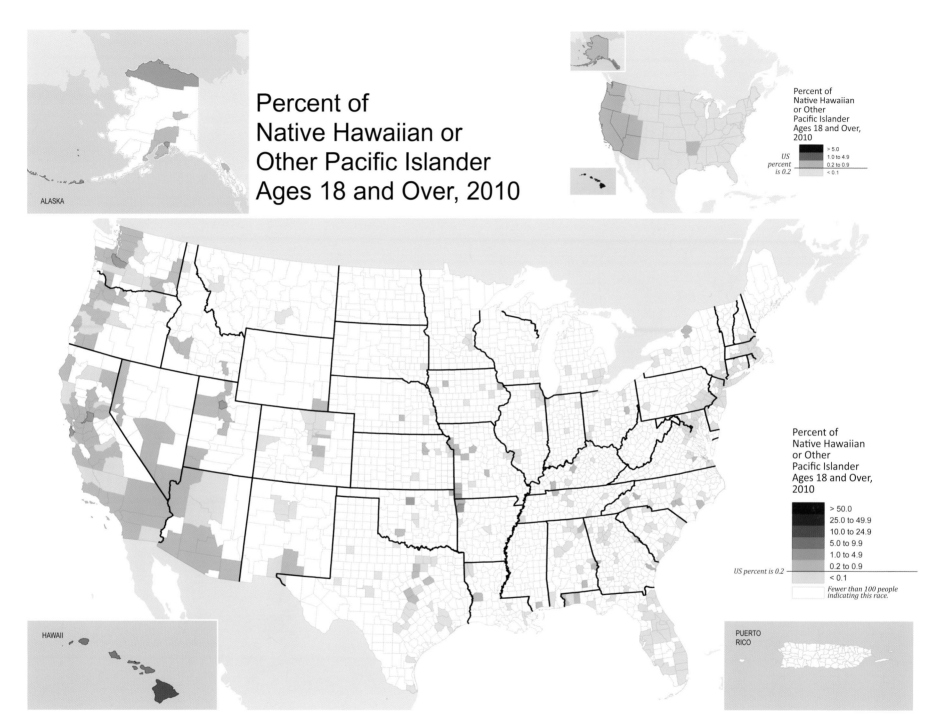

Percent of
Native Hawaiian or
Other Pacific Islander
Ages 18 and Over, 2010

Percent of
Native Hawaiian
or Other
Pacific Islander
Ages 18 and Over,
2010

US percent is 0.2

> 5.0
1.0 to 4.9
0.2 to 0.9
< 0.1

ALASKA

Percent of
Native Hawaiian
or Other
Pacific Islander
Ages 18 and Over,
2010

> 50.0
25.0 to 49.9
10.0 to 24.9
5.0 to 9.9
1.0 to 4.9
0.2 to 0.9
< 0.1

US percent is 0.2

Fewer than 100 people
indicating this race.

HAWAII

PUERTO
RICO

 These maps show the share of the population that chose only one race category. This race category can be of either Hispanic or non-Hispanic ethnicity.

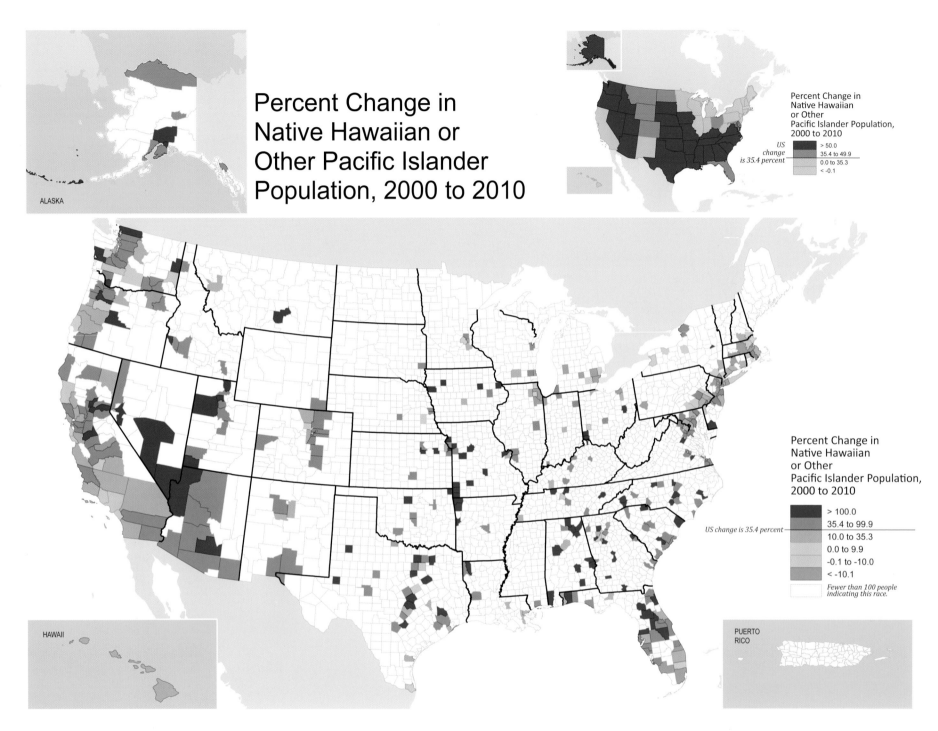

Percent Change in Native Hawaiian or Other Pacific Islander Population, 2000 to 2010

ALASKA

Percent Change in Native Hawaiian or Other Pacific Islander Population, 2000 to 2010

US change is 35.4 percent

> 50.0
35.4 to 49.9
0.0 to 35.3
< -0.1

Percent Change in Native Hawaiian or Other Pacific Islander Population, 2000 to 2010

US change is 35.4 percent

> 100.0
35.4 to 99.9
10.0 to 35.3
0.0 to 9.9
-0.1 to -10.0
< -10.1
Fewer than 100 people indicating this race.

HAWAII

PUERTO RICO

These maps show the share of the population that chose only one race category. This race category can be of either Hispanic or non-Hispanic ethnicity.

Two or More Races

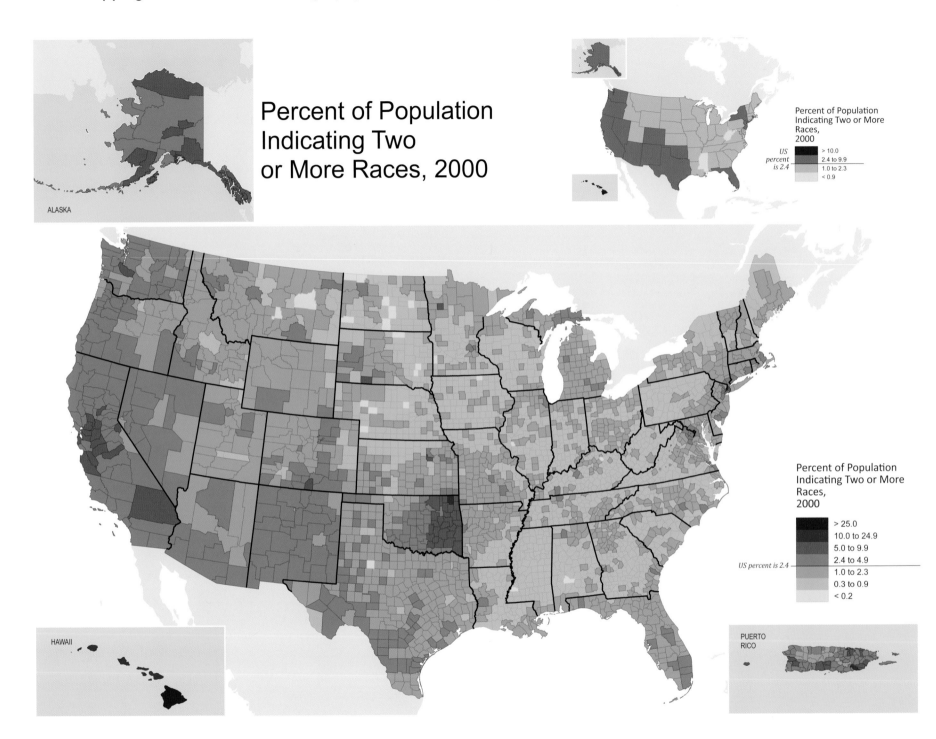

Percent of Population
Indicating Two
or More Races, 2000

ALASKA

Percent of Population
Indicating Two or More
Races,
2000

US percent is 2.4

> 10.0
2.4 to 9.9
1.0 to 2.3
< 0.9

Percent of Population
Indicating Two or More
Races,
2000

> 25.0
10.0 to 24.9
5.0 to 9.9
2.4 to 4.9
US percent is 2.4
1.0 to 2.3
0.3 to 0.9
< 0.2

HAWAII

PUERTO
RICO

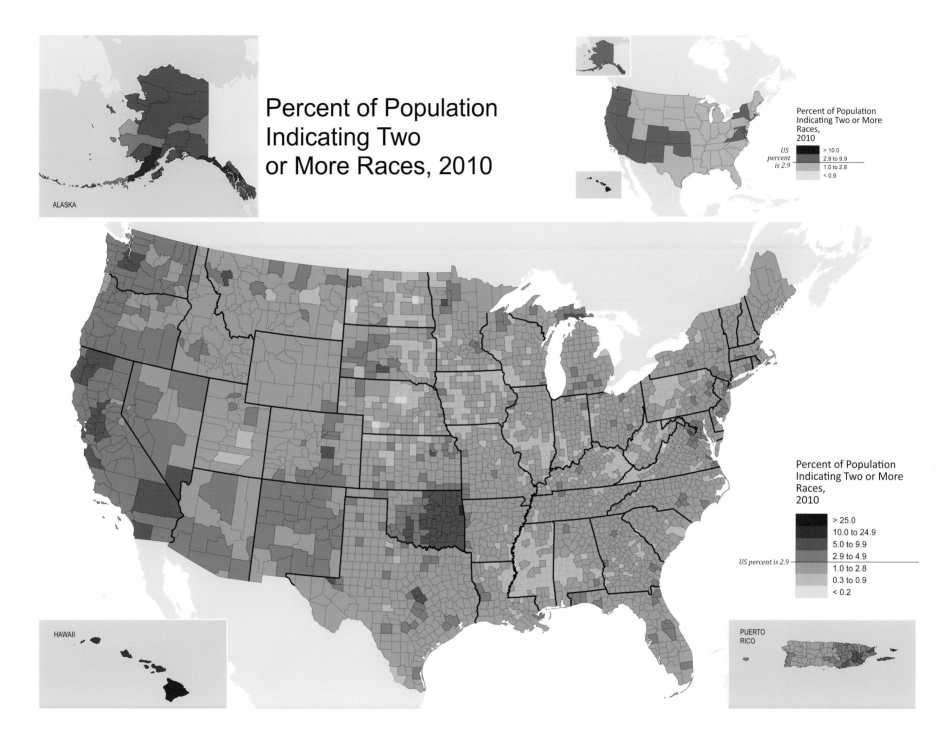

Percent of Population Indicating Two or More Races, 2010

Percent of Population Indicating Two or More Races, 2010

US percent is 2.9

> 10.0
2.9 to 9.9
1.0 to 2.8
< 0.9

ALASKA

HAWAII

PUERTO RICO

Percent of Population Indicating Two or More Races, 2010

> 25.0
10.0 to 24.9
5.0 to 9.9
2.9 to 4.9
US percent is 2.9
1.0 to 2.8
0.3 to 0.9
< 0.2

These maps show the share of the population that chose more than one race category. These race categories can be of either Hispanic or non-Hispanic ethnicity.

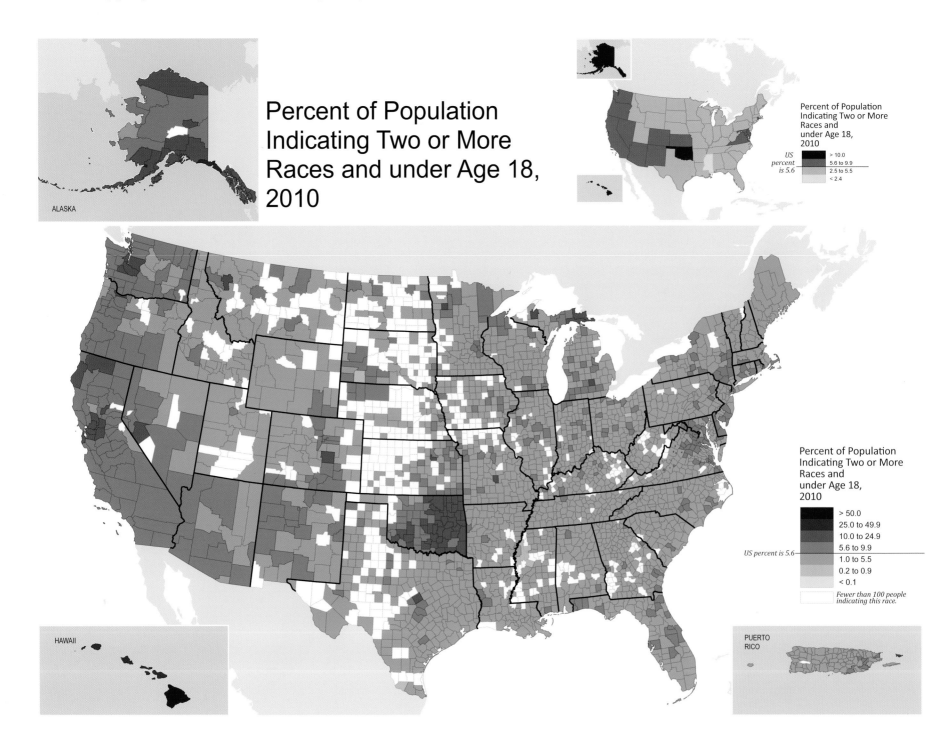

Percent of Population Indicating Two or More Races and under Age 18, 2010

ALASKA

Percent of Population Indicating Two or More Races and under Age 18, 2010

US percent is 5.6

> 10.0
5.6 to 9.9
2.5 to 5.5
< 2.4

Percent of Population Indicating Two or More Races and under Age 18, 2010

US percent is 5.6

> 50.0
25.0 to 49.9
10.0 to 24.9
5.6 to 9.9
1.0 to 5.5
0.2 to 0.9
< 0.1

Fewer than 100 people indicating this race.

HAWAII

PUERTO RICO

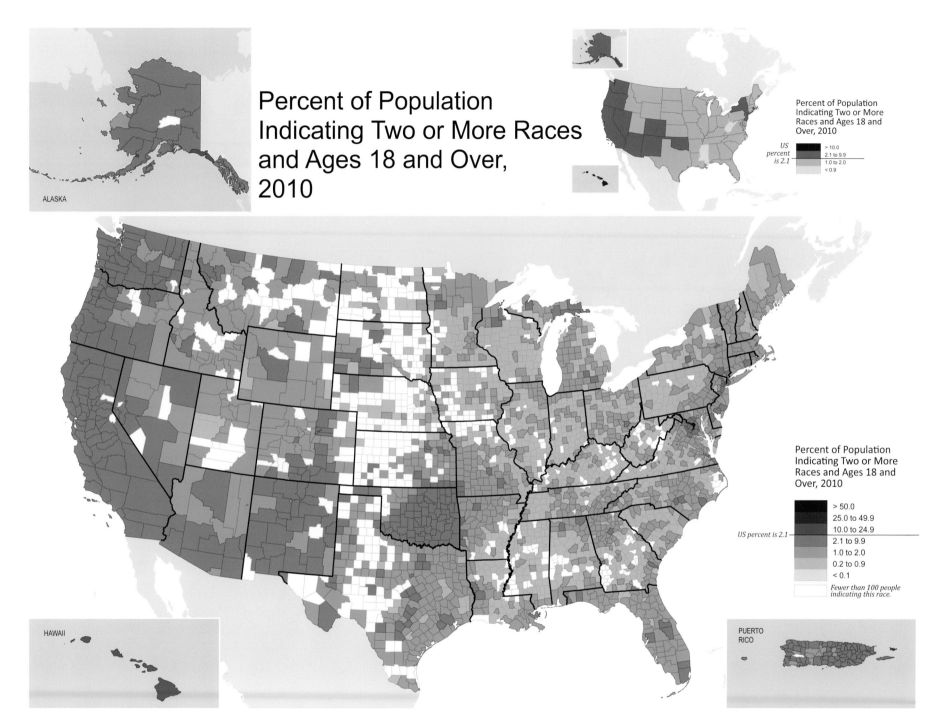

Percent of Population Indicating Two or More Races and Ages 18 and Over, 2010

ALASKA

Percent of Population Indicating Two or More Races and Ages 18 and Over, 2010

US percent is 2.1

> 10.0
2.1 to 9.9
1.0 to 2.0
< 0.9

Percent of Population Indicating Two or More Races and Ages 18 and Over, 2010

US percent is 2.1

> 50.0
25.0 to 49.9
10.0 to 24.9
2.1 to 9.9
1.0 to 2.0
0.2 to 0.9
< 0.1

Fewer than 100 people indicating this race.

HAWAII

PUERTO RICO

 These maps show the share of the population that chose more than one race category. These race categories can be of either Hispanic or non-Hispanic ethnicity.

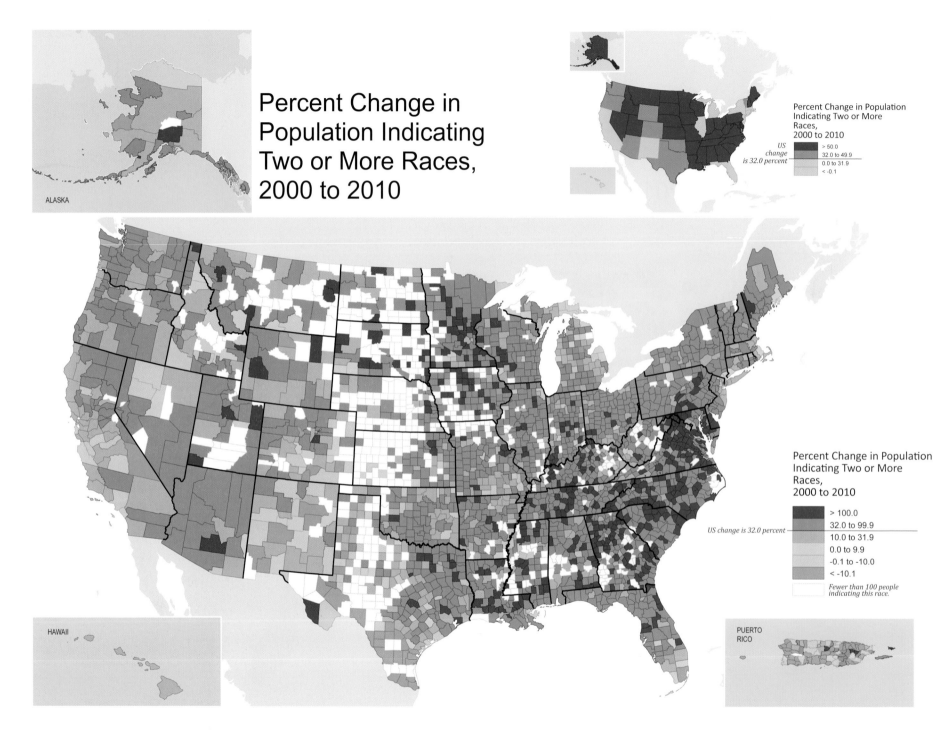

Percent Change in Population Indicating Two or More Races, 2000 to 2010

ALASKA

Percent Change in Population Indicating Two or More Races, 2000 to 2010

US change is 32.0 percent

- \> 50.0
- 32.0 to 49.9
- 0.0 to 31.9
- < -0.1

Percent Change in Population Indicating Two or More Races, 2000 to 2010

US change is 32.0 percent

- \> 100.0
- 32.0 to 99.9
- 10.0 to 31.9
- 0.0 to 9.9
- -0.1 to -10.0
- < -10.1
- *Fewer than 100 people indicating this race.*

HAWAII

PUERTO RICO

These maps show the share of the population that chose more than one race category. These race categories can be of either Hispanic or non-Hispanic ethnicity.

Hispanic or Latino Origin

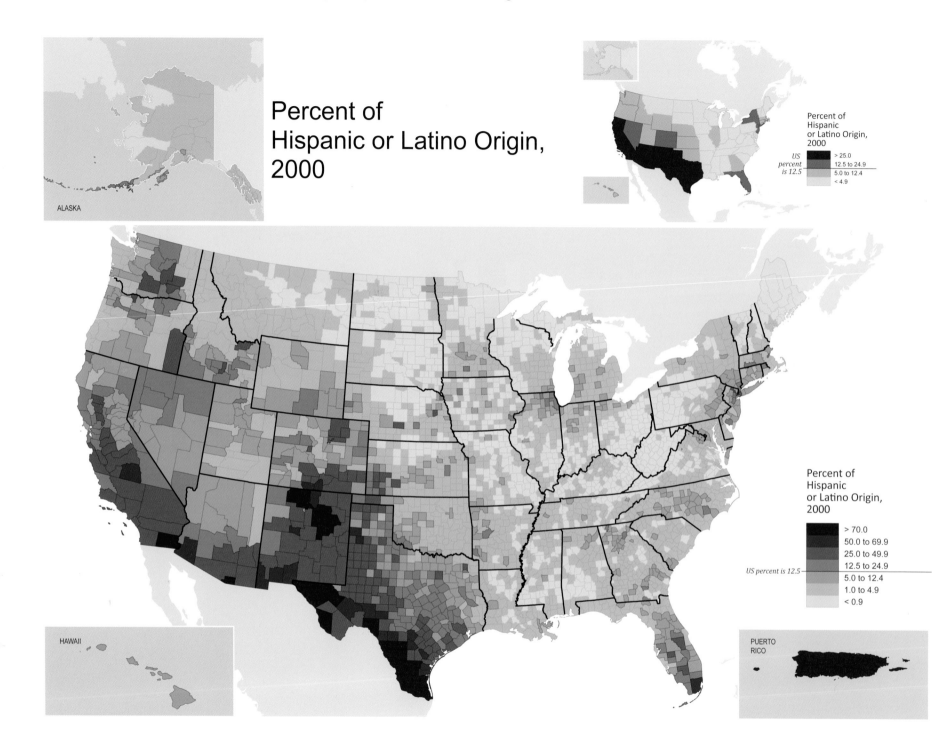

Percent of
Hispanic or Latino Origin,
2000

Percent of
Hispanic
or Latino Origin,
2000

US
percent
is 12.5

> 25.0
12.5 to 24.9
5.0 to 12.4
< 4.9

ALASKA

Percent of
Hispanic
or Latino Origin,
2000

> 70.0
50.0 to 69.9
25.0 to 49.9
12.5 to 24.9
5.0 to 12.4
1.0 to 4.9
< 0.9

US percent is 12.5

HAWAII

PUERTO
RICO

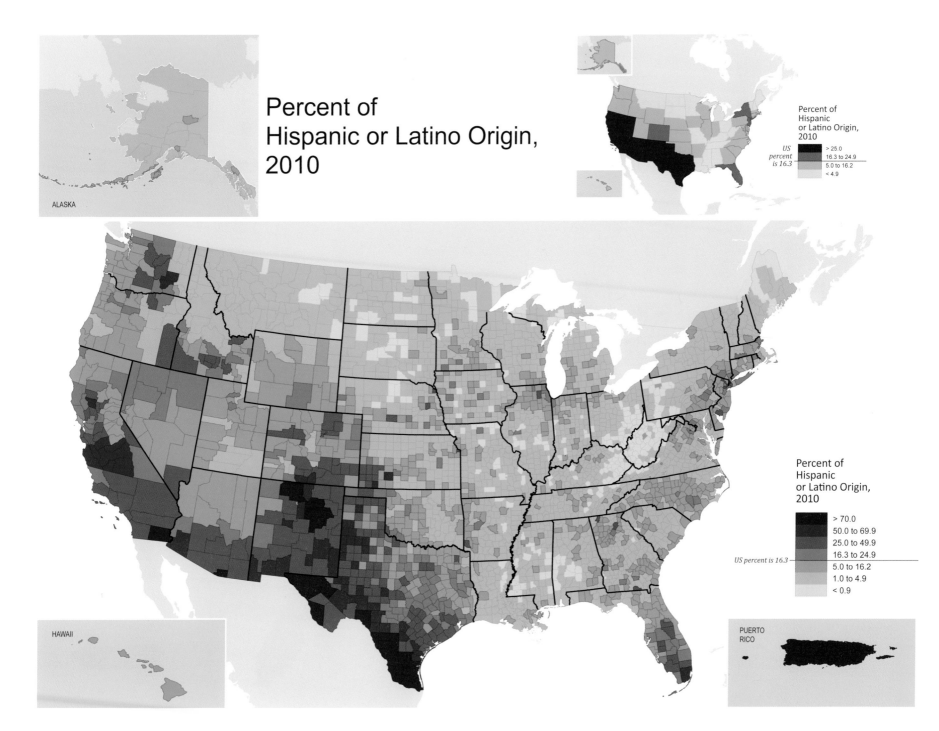

Percent of
Hispanic or Latino Origin,
2010

ALASKA

Percent of
Hispanic
or Latino Origin,
2010

US
percent
is 16.3

> 25.0
16.3 to 24.9
5.0 to 16.2
< 4.9

Percent of
Hispanic
or Latino Origin,
2010

> 70.0
50.0 to 69.9
25.0 to 49.9
16.3 to 24.9
US percent is 16.3
5.0 to 16.2
1.0 to 4.9
< 0.9

HAWAII

PUERTO
RICO

 These maps show the share of the population that chose Hispanic ethnicity. Hispanics can be of any race category.

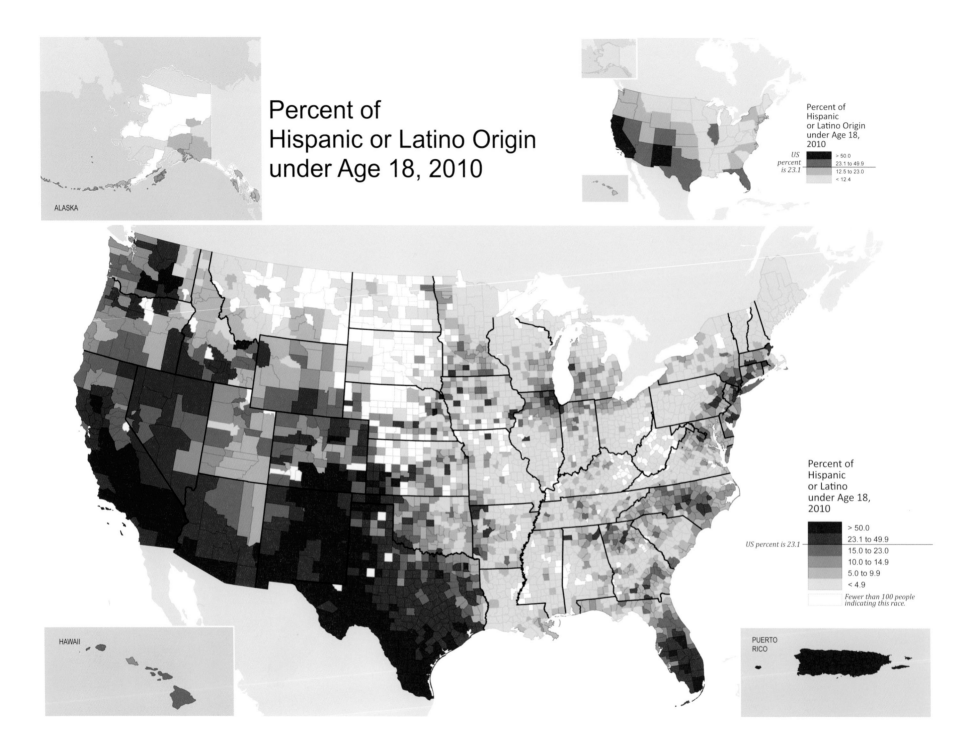

Percent of
Hispanic or Latino Origin
under Age 18, 2010

ALASKA

Percent of
Hispanic
or Latino Origin
under Age 18,
2010

US percent is 23.1

> 50.0
23.1 to 49.9
12.5 to 23.0
< 12.4

Percent of
Hispanic
or Latino
under Age 18,
2010

> 50.0
23.1 to 49.9
US percent is 23.1 — 15.0 to 23.0
10.0 to 14.9
5.0 to 9.9
< 4.9

*Fewer than 100 people
indicating this race.*

HAWAII

PUERTO
RICO

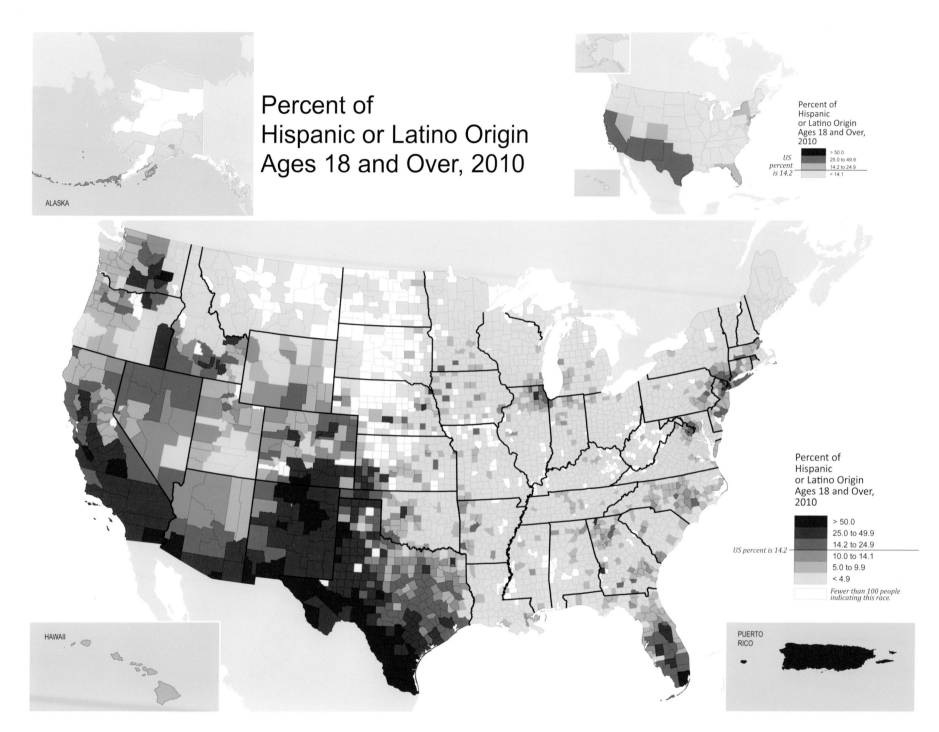

Percent of
Hispanic or Latino Origin
Ages 18 and Over, 2010

ALASKA

Percent of
Hispanic
or Latino Origin
Ages 18 and Over,
2010

US
percent
is 14.2

> 50.0
25.0 to 49.9
14.2 to 24.9
< 14.1

Percent of
Hispanic
or Latino Origin
Ages 18 and Over,
2010

US percent is 14.2

> 50.0
25.0 to 49.9
14.2 to 24.9
10.0 to 14.1
5.0 to 9.9
< 4.9
Fewer than 100 people indicating this race.

HAWAII

PUERTO
RICO

These maps show the share of the population that chose Hispanic ethnicity. Hispanics can be of any race category.

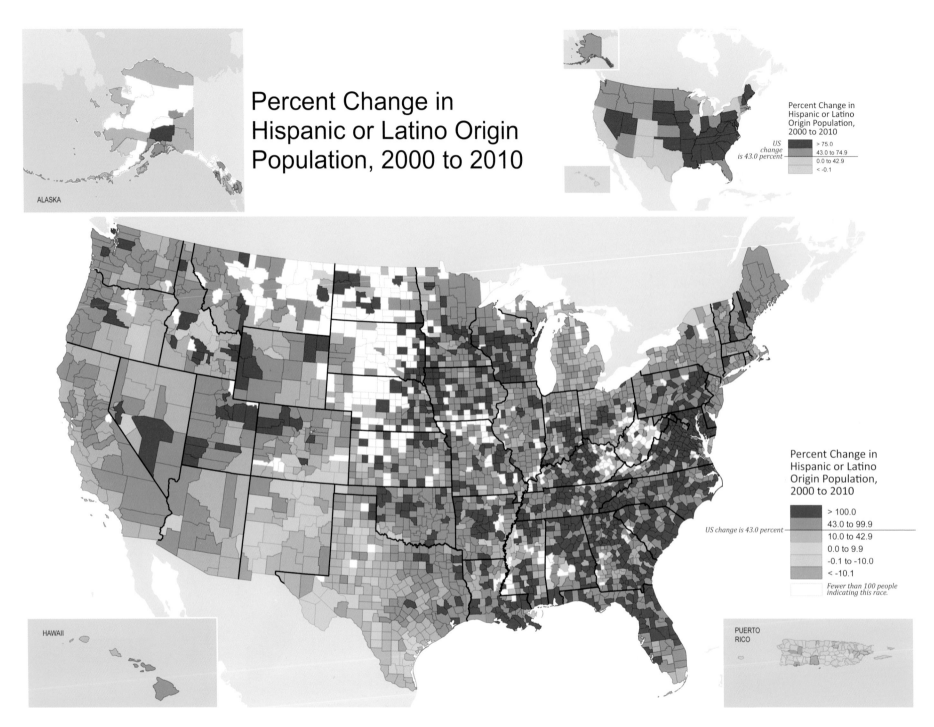

Percent Change in
Hispanic or Latino Origin
Population, 2000 to 2010

Percent Change in
Hispanic or Latino
Origin Population,
2000 to 2010

US
change
is 43.0 percent

> 75.0
43.0 to 74.9
0.0 to 42.9
< -0.1

ALASKA

Percent Change in
Hispanic or Latino
Origin Population,
2000 to 2010

US change is 43.0 percent

> 100.0
43.0 to 99.9
10.0 to 42.9
0.0 to 9.9
-0.1 to -10.0
< -10.1
*Fewer than 100 people
indicating this race.*

HAWAII

PUERTO
RICO

These maps show the share of the population that chose Hispanic ethnicity. Hispanics can be of any race category.

Data Sources

StatesLine from Data and Maps for ArcGIS 2010, courtesy of Esri, derived from Tele Atlas, US Census, Esri (Pop2010 field).

Counties courtesy of US Census.

CanadaGL from Data and Maps for ArcGIS 2010, courtesy of ArcWorld Supplement.

Russia from Data and Maps for ArcGIS 2010, courtesy of ArcWorld Supplement.

Canada/Mexico from Data and Maps for ArcGIS 2010, courtesy of ArcWorld Supplement.

PuertoRico from Data and Maps for ArcGIS 2010, courtesy of ArcWorld Supplement.

Census 2000 Data (US, County, Puerto Rico) courtesy of US Census Bureau, Census 2010 Summary File 1. Esri converted Census 2000 data into 2010 geography.

Census 2010 Data (US, County, Puerto Rico) courtesy of US Census Bureau, Census 2010 Summary File 1. Esri converted Census 2000 data into 2010 geography.

LargeCities from Data and Maps for ArcGIS 2010, courtesy of Tele Atlas.

StateLargest from Data and Maps for ArcGIS 2010, courtesy of Tele Atlas.

LargeCitiesAnno from Data and Maps for ArcGIS 2010, courtesy of Tele Atlas.

StateLargestAnno from Data and Maps for ArcGIS 2010, courtesy of Tele Atlas.

StatesAnno from Data and Maps for ArcGIS 2010, courtesy of Tele Atlas.

USHillshade from Data and Maps for ArcGIS 2009, courtesy of USGS EROS Data Center.